W9-BEG-543

The Revealing Lens

The Revealing Lens

MANKIND AND THE MICROSCOPE

Brian J. Ford

With a Foreword by
PROFESSOR JOHN BUNYAN

Harrap London

First published in Great Britain 1973
by GEORGE G. HARRAP & CO. LTD
182-184 High Holborn, London WC1V 7AX

ISBN 0 245 51016 8

Composed in Monotype Times New Roman 327 and Monotype Univers Bold 693
Typeset by Gloucester Typesetting Co. Ltd, Gloucester
Printed by Redwood Press Ltd., Trowbridge, Wiltshire

MADE IN GREAT BRITAIN

Foreword

BRIAN FORD is a first-class man of science who has not been 'blinded by the marvels of science'. His writings show a wide knowledge of the history of science; he weighs up the available facts and tells us about those who anticipated the discoveries so often attributed to other men. Some of the most persistent popular myths are exploded, and recognition is given to many of those who deserve it.

This book does not follow the usual pattern of works on the microscope. Optical theory, detailed description of instruments and accessories, and accounts of what can be seen are given their proper place. Henry Baker, in his charming eighteenth-century book on the microscope, says, 'There are many pretty little contrivances sold at the shops for the viewing of small objects which are entertaining as far as can be expected from them, but to enumerate them all would be tedious talk'. To concentrate on the instrument and in the process to neglect the personalities has often caused difficulties since Baker's time. But Brian Ford has not made this mistake.

The writer gives us valuable and interesting insights into discoveries with the microscope. He describes early work which contributed to better-known discoveries, and reveals many facts hitherto unknown to most readers.

The clear and refreshing style and wide range of this book will give the reader, both scientific and lay, a great deal of useful information, and at the same time much pleasure.

John Bunyan
Past President, Royal Microscopical Society and
Visiting Professor, University of Galveston, Texas

Contents

Illustrations

Introduction

MICROSCOPES are everywhere. In toyshop windows, supermarkets, even in our local petrol station there is a kit. Microscopes—as everyone knows—are the vital tool of the industrial scientist, the detective, the doctor; never was there a more definitive hallmark of scientific endeavour than this.

Yet what is a microscope? What are its limitations and its promise for the future? Concepts of magnification—and the ultimate limits that are involved—are more difficult to grasp, for, though the microscope has existed for three and a half centuries, hardly any of the general public are conversant with its operation or with the structures it reveals.

We all wash our hands before meal-times with a near-religious zeal; even a little child has heard of 'germs'. But do we all know what germs are like, how they live, or how they do what they do? Since magnifying glasses are so ubiquitous (they are found even in Christmas crackers these days) everyone is familiar with optical magnification. But is everyone equally familiar with the same lenses in the more esoteric context of microscopy? And do people turn their lenses on everyday structures to reveal hidden details of intricacy and beauty in the same way as they use binoculars to watch for craters on the moon?

No—there is almost a blind spot in our everyday awareness, an unspoken ignorance of these smaller, all-important structures that surround us. As we shall see, a similar short-sightedness has affected microscopic science in the past, too, and afflicts us still today.

So let us examine the development of the instrument, and the effects that its revelations have had on mankind. This is not just a history of the microscope, then; it is an examination of the dawn of microscopic consciousness and its implications for science, for scientific discovery, and for civilization.

Most retrospective works—histories of science and the like—are not quite what they purport to be. The cover may suggest that the reader will find therein an account of the development of the field, but that, clearly, is impossible. No volume, indeed no number of volumes, could ever encompass the growth and unravelling of a science's many facets, and it is useful to bear in mind, when reading any of these works, that the contents are inevitably selective.

This statement is not as self-evident as may first appear, for it admits a corollary that is rarely mentioned—what are the criteria for the selection? In short what does one really find between the covers of the volume—and why is it there?

All too often the events, people and places that are described are those that are best known—that is, those that have received the greatest degree of publicity for one reason or another—rather than those that are essentially the most significant. The reasons why they became candidates for such treatment are many; they include public interest in and acclaim for an item of news, of course, but this often has no relevance whatever to the true merit of the contribution itself. Often the individual who is generally believed by the public to have spearheaded some new departure or other is actually the mere figurehead of a move begun by the labours of some unsung, perhaps forgotten, pioneer. Many have been those tireless workers, striving towards the unknown, who have failed to have their ideas accepted or understood in their lifetimes (and often thereafter, too) because the mass view at that time was unprepared, the conceptual ground infertile; yet from this work have spread ripples through the pool of informed opinion which have reorientated society's approach to the subject so that the later arrival—who may merely repeat and extend the earlier work—is seen as the true pioneer. In essence it was the fortunate man who was, as one says, 'in the right place at the right time'.

In this way the orthodox picture of the history of the microscope is largely erroneous. In revealing the existence and, to some extent, the nature of the minute living things that surround mankind, the microscope may be seen as a social tool of the first order—an instrument of insight into the nature of our species and a unique key to the environment in which we see ourselves. It was this single device which through the overthrow

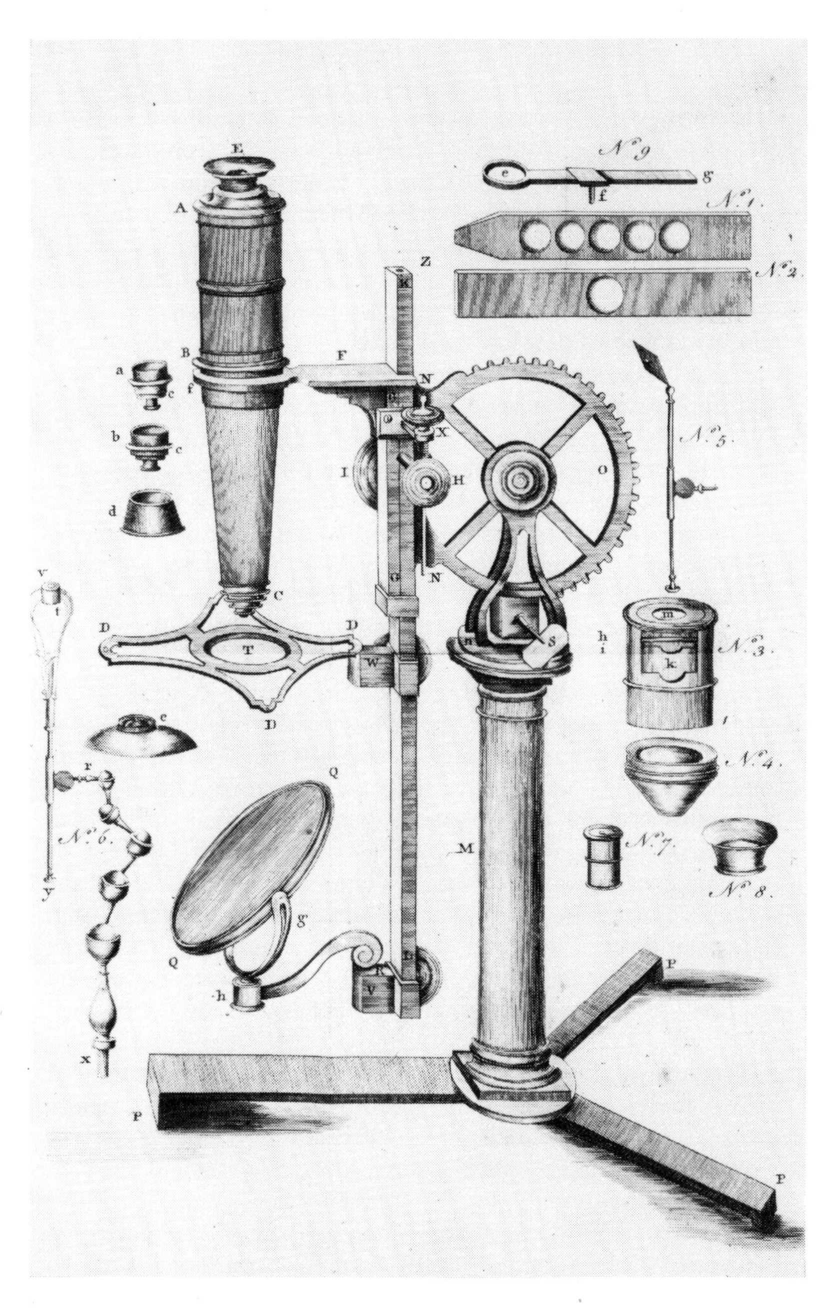

Adams's 'Variable' microscope is shown in this engraving from Hill's *Construction of Timber* (1770).

of doctrines such as that of spontaneous generation caused a complete revision of thought in the widest field, from the classroom to the hierarchy of the Church; it was the microscope that did more than anything else to reveal the true origin of infectious diseases and to explain the functioning of man himself. And the individuals who are so legendary—Pasteur, Lister, Fleming and the rest—are no more than tangible baubles of memory in a contorted web of developing thought.

There had to be social consequences of the research, and there had to be social pressures to direct it. Above all stood the desire of scientists to find out more about the new world they could perceive, to find out how to handle the instrument, and, later, the organisms it revealed. That this work led inexorably towards the attempts of later workers to control harmful species of microorganism is undeniable—but we would do well to bear in mind the secondary, almost incidental, nature of these developments. The research workers were not aiming towards some lofty ideal that later men would realize. They were determined to promulgate their own ideas, to satisfy their own curiosity (or ego), to probe deeper, deeper, ever deeper; and the overall picture that can be so clearly seen retrospectively is an historical mnemonic, and not a logical pathway to progress. To pretend otherwise is as fatuous as claiming that the dropping of the bomb in 1945 was carried out as a necessary precursor to the development of nuclear power stations, or that the laser was invented to aid the surveyor; these are later rationalizations of a self-generating, self-sustaining race towards the unknown. Does anyone seriously believe that the housewife's silicone-treated saucepans were really made better, sooner or cheaper because of the Americans' scramble to the moon?

The recurrent motivation of man's desires for advancement on the individual scale, rather than as part of a grand design, implies a certain constancy in the processes of research. And, as we shall see in this volume, there is evidence that there is as much hasty condemnation of new developments as there was a century or two ago, contradicting the notions that research is more purposive and more scientific these days and that traditionalism is no longer a hindrance to progress.

If the motives were clinical, objective, noble in some way, then science would be more sensible in its progress and a good deal

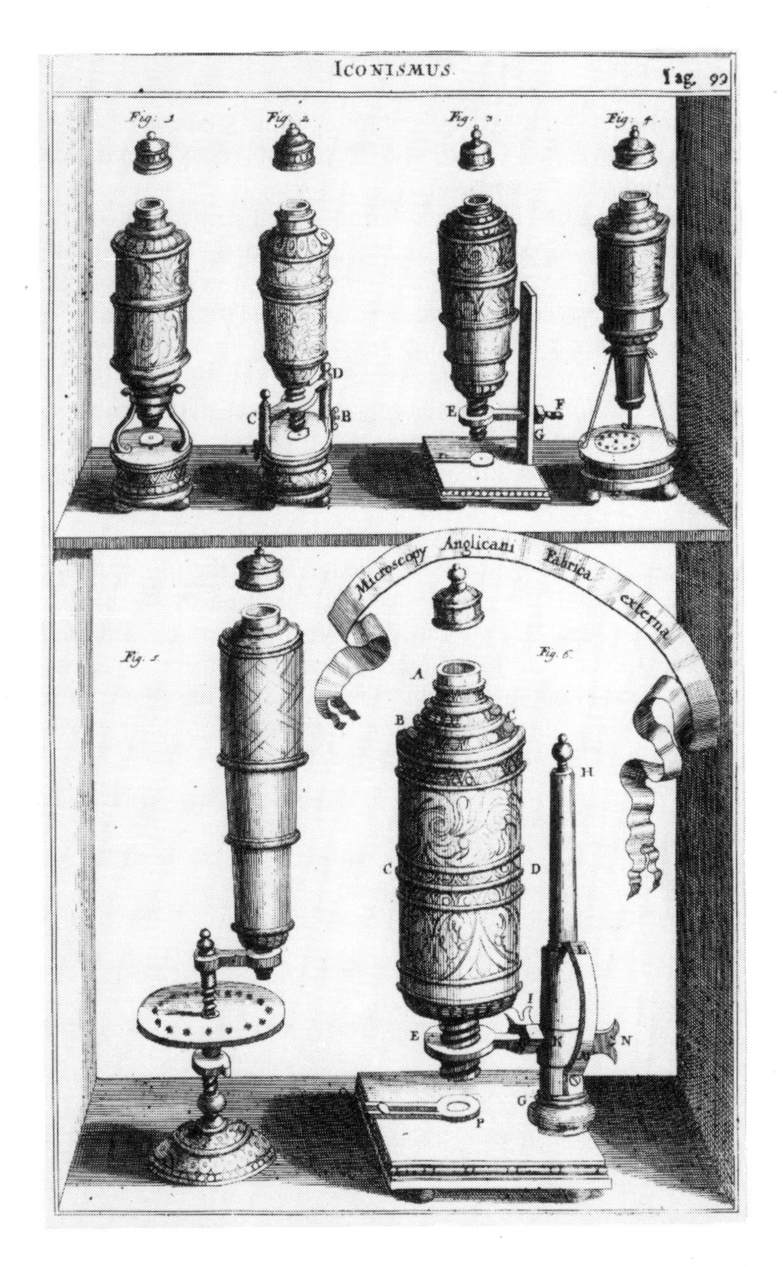

A selection of pioneer English microscopes are shown in this plate from Zahn's *Oculus Artificialis* (1685). Note the Hooke type *(lower right)* which can be compared with the illustration on page 35.

more humanitarian in its implications. But no; the conservatism of the sixteenth century is still with us since its roots lie in human nature—and that remains.

This book is a series of ideas. It discusses the way in which the microscope developed from a simple toy to become a fundamental item of the scientist's armoury; it shows some of the personal and social pressures that mould the pathways of research; it mirrors the failings inherent in the human psyche that are stumbling-blocks today as they were in the time of Galileo; it outlines some of the social pressures that were set up by a relatively simply observation—and postulates that similar pressures from earlier discoveries led science still further along its way.

Above all we may discern, through this single theme, the personal traits that nudge new knowledge from the universe and how, all too often, they have become obscured by regimented and artificial analyses in the past. We may see by mirroring the stories of the past in the light of today's topical events, how the same factors operate in our current system of science; how far we have come through one simple invention, in other words, and how much we yet have to learn.

1 Dawn of a New Science

THERE have always been lenses. Any transparent bead, a pellet of amber from the shores of the Baltic or a misty pebble of glass from a primitive furnace, is a lens of sorts. The first treatise on lens manufacture and optics was Alhazen's *Optical Thesaurus*, published in 1038, which laid the basis for European medieval optics. But the knowledge of what one might call empirical optics was well known to the ancients. Three-thousand-year-old convex ground structures made from crystal have been found, Pliny wrote of the focussing effects of a vessel of water acting as a lens on the sun's rays and hence a source of concentrated heat, and Seneca wrote that objects appeared larger when viewed through a glass of water—but seems to have concluded that it was the liquid which had powers of magnification, rather than the lens-like shape it was taking.

Several early writers (such as Vitello) described optical theory but based their work entirely on an unsubtly plagiarized interpretation of Alhazen, but it was Bacon who took the next significant step forward when he wrote of the use of lenses as magnifiers. He described their employment by those with failing sight as an aid to reading and mentioned the observation, through positive lenses, of 'many things of the like sort which persons unacquainted with these things would refuse to believe'. From these words he has been taken by many historians to be the possible true father of microscopy but it seems arguable that he would have suggested more of the nature or existence of microscopic organisms had he seen them. It is perhaps more rational to assume that he was speculating on the future, rather than recording his experiments.

It was early in the fourteenth century that lenses were first made into spectacles for general use, and thus entered the practical plane of mankind's endeavour. So lenses had, after being for centuries no more than toys or curiosities, at last arrived. Exactly when these glasses were invented, or by whom, cannot be stated with certainty. Once man had learned of the benefits that derived from the use of a magnifying glass by the aged it was obvious that he would toy with various types of supporting framework; and once he had become aware of the improvements to vision that could result from the use of one lens it is clearly obvious that he might speculate on the use of two, one for each eye. It did not call for any one genius to see the possibilities, which were more likely realized by the simultaneous curiosity of many intelligent men toying with lenses as a present-day child plays with magnets. The attractions, the benefits, the potentialities are self-evident in this sense. Some say that spectacles were invented in Florence about 1280, others that they were discovered by della Spina, of Pisa, about the same time; and one of his fellow-monks wrote in 1305 that he had known the inventor of spectacles, who had made the discovery 'twenty years before this time'. And meanwhile we are left with the statement on the tombstone of Amati, a Florentine nobleman who died in 1317, that he had invented spectacles—but had kept secret the development from all but his closest friends.

For the reasons I have outlined, it is irrational to ascribe the invention to anyone. We know of these men by name through the relatively chance occurrence of an inscription on a gravestone, or a passage in an old manuscript; but knowledge of this kind does not proliferate in such a deliberate manner. Even today we have seen examples of men heralded as the 'originators' of some move or other who were in essence single examples from a whole field of conscious development—but who were singled out for special attention because of some accident of topicality; and once that has happened the wider public interest that results may secondarily create the impression of a new development initiated by one person. Thus in the modern world one has 'originators' of jazz, pop, long hair and women's liberation—people who were not 'spearheads' at all in reality, but merely selected examples from a whole new wave.

Given the state of the development of lens grinding in the late thirteenth century it is likely that myriad experimenters were trying them out, first this way, then that, in the hope of seeing something new by way of a trick of the light or some means of

Descartes's design for a microscope, from *La Dioptrique* (1637). It was intended to be a small instrument; the observer—having been indicated roughly in the original drawing—has been inserted by the engraver, and gave rise for some time to the fanciful idea that Descartes had built the tallest microscope in history.

aiding the poor-sighted. It would be grossly insulting to the intelligence and innate curiosity of man if we did not assume this to be so—in which case there must be numbers of lens makers and primitive spectacle pioneers disintegrating in the soil of Europe who did not have any occasion to be written about, whose writings are no longer extant, or who could not afford elaborate wording on their headstones. Spectacles grew out of the state of man through his personal inquisitiveness, and not through personalized genius.

The development was, therefore, virtually autogenous: it resulted from the state of technology at that time. However the next logical step, namely the siting of two lenses along the same axis, rather than side-by-side, took longer. There cannot be any doubt that workers at that time held lenses in front of each other, even if only to see whether they supplemented each other's power—or idly to find out if two would be twice as beneficial as one. Bacon wrote of making 'far distant things appear very near . . . the stars descend lower in appearance and to be visible over the heads' and—given a few lenses to play with—any intelligent enquirer at that time must have experimented with two-lens systems.

But at that time the imperfections in the individual lenses must have been the limiting factor. In using a lens as an eyepiece, with another as objective, the user would find that the imperfections of the objective lens were so magnified by the other as to render it unusable. Experiments with an ox-eye lens from the front of a modern torch, or using a cheap plastic lens from a child's cracker at Christmas time, may well give us an idea of the difficulties these early workers must have experienced. However, as lenses improved, so men were more successful in utilizing them to penetrate distances. And so the telescope began to emerge.

There were many pre-Galilean experimenters who built telescopes. Most, no doubt, were obscure men who may have been too frightened of religious repercussions to announce their findings; but coincident with Galileo's experiments of 1609 (in which he built a telescope with a convex objective and a concave eyepiece lens) were similar experiments carried out by Hans Lippershey of Zeeland (Holland) which were reported to have been the result of an 'accidental' observation during some play with lenses. It has been recorded that he produced a binocular telescope in December 1608 for the military use of the Dutch authorities, and earlier that year a native of Alkmaar, Holland, had submitted a patent claim for an instrument which could 'bring far objects nearer to the eye'. He was James Metius, regarded by many historians as the true 'father of the telescope'. Yet far earlier, in the mid-1500s, others were probably making the same primitive observations but using a curved mirror as objective instead (thereby eliminating one important source of aberration). In *Pantometria*, published in 1571, Digges writes that his father, by using 'proportional glasses situate at convenient angles', had been able to 'declare'—over a distance of seven miles—'what hath been doone at that instante in private

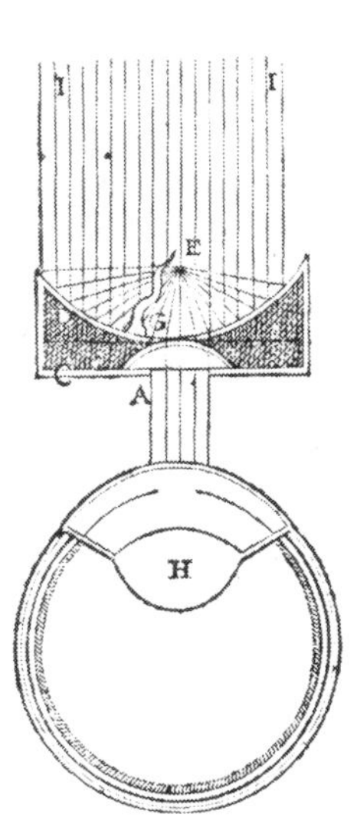

A design for a simple microscope by Descartes (1637). The observer was allowed to remain a respectable size on this occasion. (Compare with the illustration on page 19.)

places'. He had also, according to the son's reports, read out the details on coins and medallions thrown on the grass some distance away by using the apparatus. This would seem to be the earliest extant recording of such a device.

The development of simple telescopes is very pertinent to our theme, since as every schoolboy knows this instrument inverted can be used as a microscope (there are toys on sale which are described as a combined telescope and microscope; it depends through which end the observer looks). Galileo, in this way, became a pioneer of the use of the microscope when he inverted one of his astronomical telescopes and peered at insects with, according to the records, considerable awe. But a score or more years before, a Frankfurt entomologist named Hüfnagel had used a form of microscope in the preparation of copperplate illustrations for a work on insects (according to Bonanni's *Micrographia Curiosa* published in 1691) and early in the 1600s (exactly when is obscure) the Janssen brothers of Middelburg, Holland, had apparently constructed microscopes in the form of a lacquered brass tube with a lens at each end, apparently held by hand in the manner of a telescope. But still, as the first of the observing microbiologists was to prove, it was more effective to use a single short-focus lens rather than a compound instrument.

In any event lens technology had by now advanced to the state where usable results were obtained and the art of the lens grinder had spread far and wide. As we shall see, the process formed a hobby for many men initially employed in other fields. Clearly in practical terms microscopy was now feasible—a whole new universe awaited discovery and observation.

Certainly by the end of the first quarter of the seventeenth century there had been many examples of primitive microscopes made in Europe and also in England. Within the next quarter-century or so microscopes were being manufactured on a commercial scale and, as curiosities, they were owned by noblemen in growing numbers.

So much for the practical infrastructure of the new discipline. But what of its conceptual aspects? On the face of it little advance had been made. But as men toy with objects so they will idly manipulate ideas, and it is interesting to see, in ancient scientific literature, how this form of speculation threw up notions that anticipated later developments. There is, and we must remain cognizant of this fact, a tendency for writers of today's era to quote such examples as though they were always the work of genius which existed 'before its time'. Thus, with the handy benefits of hindsight, we select those hypotheses which have been confirmed by later work and herald them as far-sighted, whilst condemning opposing views as cranky and ill-founded.

This idea is itself unbalanced. We know (or, rather, *think* we know) that our present-day ideas about combustion, life, the nature of light and the rest are 'correct' because of our inordinately greater experimental knowledge and insight, and therefore such antiquated ideas as the phlogiston theory of burning, or the spontaneous generation of life, are seen too easily as false and irrelevant notions perpetuated by the ignorant—or, at best, by the innocently naïve. But at the time contemporary with the derivation of those views so little was known about the processes themselves that *any* approach which fitted the available data was probably as valid as any other. We find that those adherents to microbiological theories who had their work confirmed by more extensive investigations were later shown to have been on the right track; whereas the spontaneous-generationists, as we may call them, were not.

But it would be erroneous to place overmuch emphasis on these retrospective assessments. One must remain aware of the historical context in which any new hypothesis was advanced. Not only this, but it is equally salutary to attempt a consideration of the factors which influence the acceptance of any revolutionary thesis. It is not enough for it to be sound: more than this, it is not *necessary* for it to be so. What matters more than anything else is that the new work should be assimilable, and presented in a form acceptable to the recipient (the public, say, or the specialist community). But as with any new idea the package must be acceptable, inviting; as we shall see later in this book there have been examples of men who made profound developments in thought but who, through a failure to project them positively, were neglected. There are many cases in recent history—Hitler, the Kray brothers, McCarthy—where the most untenable and in many instances cruelly inhuman behaviour was granted local acceptance through its mode of presentation: forceful, convincing and confident. There have been many popular figures in science, as we shall see, who have had this charismatic quality. Sadly, mankind is willing to give credence to too many dogmatic ideals if the presentation is powerful—they may be accepted (and this is a cardinal failure in society) without any serious attempt to question their rationale.

Conversely, a failure to present a thesis in an assimilable form may equally lead to its being ignored. Such was the fate of Fracastoro (widely known in the literature as Fracastorius) and his brilliant speculations on the nature of disease. He wrote of contagion, of 'germs', of infectious illnesses, as a concise and workable thesis well over three centuries before Pasteur's work on the subject. But Fracastoro's book on the subject, *De Contagione* (it was issued as three volumes), was written in a condensed and idiosyncratic form of technical Latin and so this work and its important new concepts did little to perpetuate microbial theory. It was, instead, ignored and largely forgotten about by those who had seen it.

Of course, this was not the first occasion on which the existence of microscopical entities had been postulated as the arbiters of disease; Varro wrote of this possibility in the first century B.C. And the notion of contagion goes back very much further.

The ancient Egyptians seem to have based some of their formalized traditions on a vague awareness of contagion, and in the book of Leviticus there is a long dissertation on the diagnosis of leprosy and the pronouncement of sufferers 'unclean'. It is noteworthy that—as Marx has pointed out—there was a public awareness of contagion before the idea ever became current amongst medical men—at that time some specialist opinion generally regarded the public as gullible to believe in such nonsense!

Most early theories of disease tended to have a strong supernatural element (it was said to be due to eclipses, comets, shooting-stars) or a moralistic one (the result of adultery, blasphemy or religious neglect). The Hippocratic concept fell some way between the two stools of supernatural influence *versus* contagion by assuming that bad air gave rise to illnesses—and the term 'malaria' is of course a literal translation of this presumptive aetiology. But during the Middle Ages the poor conditions of sanitation and hygiene, coupled with the ensuing ravages of pestilence and plague, gave a growing pragmatic insight into the contagious nature of many illnesses, borne out by the account of the 1348 epidemic in Florence (written by Boccaccio and published in *Decameron*) in which there is a sound account of the transmission of illnesses *via* contaminated clothing, coupled with the assertion that the outbreak was due to corrective measures invoked by God. A strange blend, but an interesting view of the state of opinion at that time.

Though it must have become increasingly widespread for ordinary people to see contagion at work, there was but little tangible progress until Fracastoro set his mind to work on the matter. No-one is quite sure when he was born; it is assumed that at his death in Bologna on 6th August 1553 he was about seventy-five years of age. Without question he was a very considerable genius; he came under the influence of many great men including Copernicus (Koppernigk) the astronomer and Pomponazzi, who was convinced that the explanations of events were physical, rather than supernatural, in nature—this latter philosophy must have been a significant element in Fracastoro's work on the origins of disease.

He studied widely in the physical and natural sciences, was a poet of distinction, and after a period as a lecturer and demonstrator in anatomy at Padua University he became the focus of many noblemen who travelled considerable distances to ask his advice on medical matters. At the same time he wrote his epic narrative poem *Syphilis sive Morbus Gallicus* which describes a shepherd who contracts syphilis. As a result of this popular work the word universally replaced the term 'the French Disease'—although some students of the period have suggested that the word 'syphilis' was already current as a colloquial expression in Fracastoro's time and that the name of the shepherd—Syphilus—was therefore a somewhat jocular play on words (contrary to the popular view, which suggests that the term syphilis originated from the name of the poem's shepherd hero). The poem was published in 1530 and the dedication was to Cardinal Bembo—himself a poet of sorts, but described as the writer of popular verse 'as obscene as any in literature'. The literary verve of Fracastoro's poem was such as to make it one of the most popular works of its day, putting even Bembo's dirty poems into the shade.

Political and student unrest at this time in the universities of northern Italy had led to the closure of some colleges. Fracastoro, who was working quietly at a villa on the shores of Lake Garda, paid infrequent visits to Verona when the political unrest had temporarily subsided. During this time he reflected on disease and its nature, the notes forming the outline of *De Contagione*, which he published in 1546 at Venice. The following year he joined the Council of Trent as medical adviser, and moved to Bologna (where he later died of a stroke) when typhus outbreaks forced the Council to reconvene there.

Such was the superstitious nature of scientific beliefs at that time that even Fracastoro wrote of conjugations of planets, rains followed by drought and other 'premonitory signs' as being amongst the signs of contagion, and he seemed to recognize an almost conscious malevolence in the purposes of disease 'germs' —'they have . . . a spiritual antipathy to the animal organism,' he wrote.

His descriptions of diseases such as foot-and-mouth and typhus are remarkable, as was his allusion to the infectious nature of consumption (pulmonary tuberculosis) whose contagion, he thought, could remain for two years in contaminated clothing. Rabies, he recognized, was transmitted only by the bite of a rabid dog and he noted the long incubation period (up to eight months, in one case he cites) that may be found. But his most important single contribution was his use of the term 'seminaria' to describe the disease-causing agency. The term is untranslatable into English; seminal fluid has the connotations of a liquid containing vital living entities, and the term 'germ' is one of the former equivalents of 'seminaria' in English—though not as we use the word today. Current English has tended to define germs as essentially harmful or disease-causing micro-organisms, but the earlier meaning was more general, and meant the centre of reproductive or proliferative activities of living organisms or vital processes. Modern relics include 'germination' and, in some older textbooks 'germ cells' (for sexual cells); some still refer to the father's 'germ' fertilizing the mother's 'seed' in human reproduction.

Thus we must recognize that the use of the word 'germ' to connote only harmful entities is a comparatively recent convention. In speaking of 'seminaria' ('germs' in the old sense) Fracastoro was alluding to his belief that diseases were due to minute, unseen, virulent living particles which had the power to induce illness. That view, arising when it did, was a singularly prophetic piece of speculation which might well have paved the way to the dawn of microbiology.

But this did not occur. Fracastoro's idiosyncratic writings did not capture the audience they deserved; they gained some currency amongst philosophers who knew his work well enough to pursue their readings of *De Contagione*, but within fifty years all his new ideas had fallen into neglect. The pebble had been too small: the ripples died quickly and quietly away. Indeed as Bulloch points out in his excellent *History of Bacteriology* William Harvey wrote a letter which showed how little effect Fracastoro's work had had; Harvey, says Bulloch, wrote to the Florentine John Nardi that he found it difficult to see how contagion 'long lurking in such things' as infected clothing, etc., could after a prolonged period 'produce its like in another body'. And this letter was dated 1653—one century exactly after Fracastoro died.

Why was this vital period of one hundred years of potential progress lost? If Fracastoro's ingenious theories had been known to the manufacturers of microscopes—then being used as nothing more than toy magnifiers—microbiology might have arisen long before it did. The coincidence of the technological development (i.e., the existence of microscopes in the first half of the seventeenth century) with the conceptual step taken by Fracastoro's writings could have thrown up a whole range of pioneer microscopists but, as it was, most of his ideas had to be independently rediscovered—sometimes, as we shall see, time and time again—before they gained acceptance.

One of the most important reasons underlying the neglect of his work was the cumbersome and unassimilable manner in which it was presented. This would not have precluded the possibility of a different investigator carrying on the project but of course we must bear in mind the very powerful strictures placed on this form of uninhibited conceptual advance by the Roman Catholic church at that time. The Inquisition was still powerful and continued to persecute heretics actively; as a result it was not only more compatible with current views, but it was more prudent too, for scientific pioneers to play down any naturalistic emphasis of their work. A singular example of the supernatural factors that were believed to influence the course of diseases is to be found in the account of Wm. Boghurst, an apothecary, of the great plague of London in 1665. It was due, according to this writer (so far-sighted and objective in many other ways) to a 'subtle, perculiar, insinuating, venemous, deleterious exhalation . . . sometymes immediately aggressing apt bodys'—a reiteration of doctrines thousands of years old. Though one could hardly condemn Boghurst's writings for the mere omission of reference to micro-organisms, it is surprising to find him so willing at that time to opt for the purely superstitious alternative.

Apart from isolated surges in the flow of advancement (such as Fracastoro's investigations) the tide of progress was ready, by the middle of the seventeenth century, for the epic breakthrough to the world of microscopic structures. And this, surely, was the moment at which the new discipline began. It was not with the simple magnifiers (which had been available for centuries) nor did it lie in the direction of the philosophers who speculated on the nature of life and disease. The simple and fundamental reason is that, until this period, men had used lenses and microscopes merely as a means of magnifying that which was already visible to the naked eye—in essence, a stage further in the use of lenses to help the aged see printing. These pioneer scientists were not using their apparatus to discover new structures at all, but merely to facilitate their observations of conventional familiar objects. An ant, obviously, was easier to draw if magnified first by a lens; and this is why microscopes were employed at all. But what was still missing was the search for microscopical structures, micro-organisms or tiny events that could not be seen at all by the eye. The arguments about who were the first men to magnify with lenses at all is no more than historical guesswork. They were, by any standards, *not* microscopists at all.

How did it eventually begin? It is unlikely that there was a moment, or a man who first stepped on a 'ladder of discovery'. It was more likely a group of scientists who began to turn slightly towards the new direction. The first notable microscopical developments dawned on several men, who suddenly realized not merely that lenses were capable of revealing conventional structures in a state of magnification, but that they could reveal entirely new and unsuspected forms of life, undreamed-of structural configurations, strange and minuscule particles . . . in short, a hitherto unseen universe. Leeuwenhoek is the name that springs most readily to mind as the first microscopist and as we shall see his work was without question in the forefront of the new wave of investigation.

But he was, most certainly, not the first man to suddenly realize the existence of the hidden world that the lens could reveal. Leeuwenhoek's work was antedated by at least ten years in the investigations of Hooke (of 'Hooke's Law' fame). It was this Englishman's writing that first coined the term 'cell' in relation to the pore-like structures he perceived in cork on the morning of 13th April 1663—but even this epoch-making discovery has a precedent of perhaps eight years, when Athanasius Kircher, a rambling and garrulous Jesuit professor of German extraction, wrote of very small and imperceptible living bodies that were so fine they could 'insinuate themselves into the innermost fibres of clothes, ropes . . . whatever is full of pores, like wood, bones, cork. . . .' He went on to declare that 'only after the invention of the Microscope did it become known that all decomposing things swarm with an innumerable brood of worms that are invisible to the naked eye: which even I would never have believed, had I not proved it by repeated experiment over many years'. These interesting extracts appear in his book *Scrutinum Pestis* dated 1658, and it is interesting to note that he suggests *not* that he has just discovered microbes in putrefying matter, but that it *became*

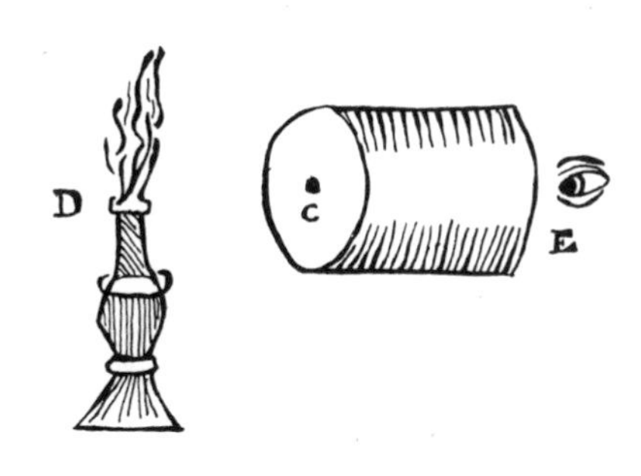

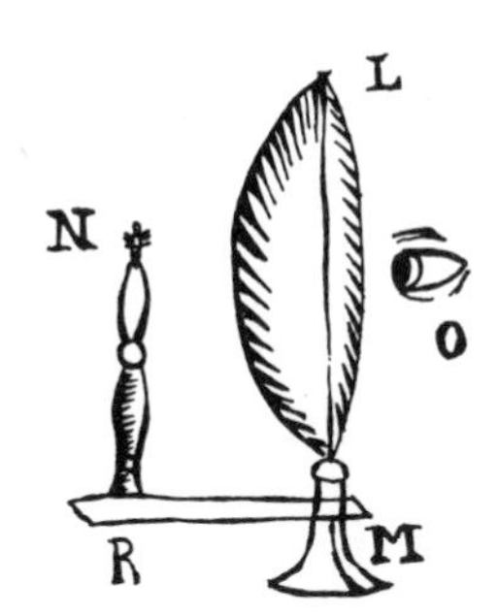

Two microscopes from Kircher's *Ars Magna Lucis* (1646). They are single-lens (i.e., simple) magnifiers which were used to observe insects. The name flea-glasses was popularly applied to them.

known—it appears to be an implication that this was the general state of knowledge at that time and suggests that others besides himself knew of the fact. We must therefore own to the fact that if he wrote (in a book published in 1658) of 'repeated experiments over many years', it is arguable that in the late 1640s or early in the next decade the existence of the previously invisible universe was beginning to make itself felt.

Though Kircher's writings are valuable for these occasional glimpses of insight and revelation, they seem in the main to be wild, speculative and conceptually incoherent. Thus he speaks of supernatural influences, he suggests that a cause of plague might be a rotting mermaid, that the devil had a hand in it too, and that unfavourable celestial conjunctions were among the more significant of the aetiological factors!

We may look back even further amongst this strange man's writings, in fact to 1646, for the most revealing clue of all. For it is in the book *Ars Magna Lucis et Umbrae*, printed in Rome in that year, that he poses some surprising rhetorical questions. 'Who would believe', he starts, 'that vinegar and milk abound with an innumerable multitude of worms?' He goes on to state that such microscopic organisms had been observed in putrid matter, and in the blood of those victims of fevers and the

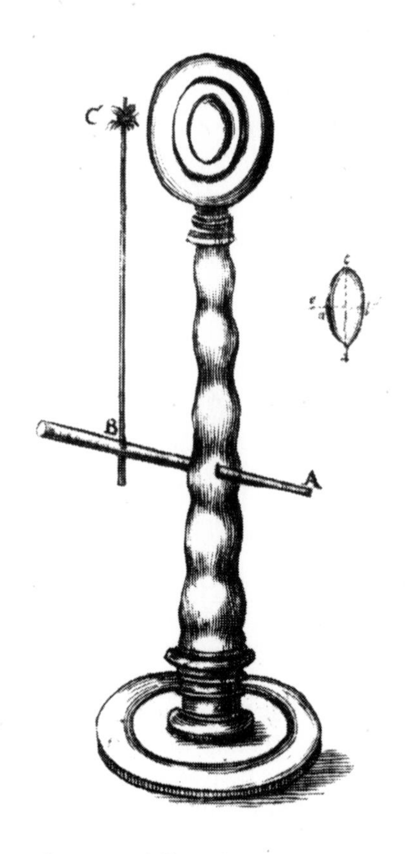

A simple flea-glass drawn by Zahn (1685). Compare with Kircher's simple magnifiers shown in the illustration on page 29.

plague, adding that plants were formed of 'different and wondrous union of filaments'. Surely here, if anywhere, is a portentous finger pointing towards the discoveries that were to come.

Many writers have ridiculed Kircher—as we have seen, not altogether without reason—but the invective has tended to get out of hand. One noted writer damns his work by stating that all he ever observed were nematodes and they, after all, could be seen by anyone with a simple hand lens today . . . but then, that is applying the condescension of hindsight. In truth most of the earliest discoveries could be so repeated. But this should not detract from the nature of the first observations, or their recording; it is easy to show a child how to cultivate bacteria or to explain the principles of survival of the fittest; none the less we would do well to honour those pioneer philosophers and scientists who found it out for themselves, for the very first time. That is scientific discovery indeed.

And did Kircher observe bacteria? Major writes that one of the remaining microscopes *could* demonstrate larger bacteria; and unless Kircher was simply applying intelligent guesswork it must remain a possibility that he may have done so. Personally I think it unlikely: but there is no evidence that would allow one to join the strong movement of historians who have, in the past, dismissed Kircher as a crank and nothing more. A crank he undoubtedly was—but he was something else besides, for his writings provide a unique insight into the state of inquiry during his life-time and (even if the accounts were vague, and probably grossly exaggerated) his are the earliest extant examples of microbiological documentation.

It was in 1661 that the first scientific microscopical discovery—at least, the first to be documented—was made. Marcello Malpighi, then a thirty-three-year-old Professor at the University of Bologna, wrote a learned communication on his investigations into the structure of lung tissue. During these he observed capillary blood flow for the first time, thereby completing the cycle necessary for Harvey's circulation theory to be confirmed. Even so Malpighi never recognized the red blood cells (erythrocytes) for what they are—not surprisingly, since there was no scientific precedent on which to base the diagnosis—and he described them as fat particles. Under a low-power lens, it must be said, erythrocytes do indeed have a certain resemblance to globules of fat and it is most unjust to imply, as some writers have done in the past, that Malpighi's identification was altogether wrong. In the terms he understood at the time it was probably as accurate as he could hope to be. Malpighi who, in the words of Rousseau, was 'le véritable créateur de la biologie microscopique', was carrying out his first work just as the Royal Society was forming in London. With the ardour of any such new, lively society they sent out communications to scientists in many fields all over Europe, and Malpighi was invited to become a corresponding member in 1667. As a result much of his later work was first published through communications to the Royal Society of London.

And it was here that further important developments were simultaneously occurring. Their nature is apparent from the fragmentary documentation that remains. Dr Henry Power, a physician, published some accounts of microscopy in 1663—the year he was elected to Fellowship of the Royal Society (thus becoming one of the first Fellows)—and correspondence written by him in 1662 and preserved in the British Museum suggests that the field-lens (i.e., an additional lens fitted to enlarge the final field of view) might be omitted from the design of certain microscopes with benefit. The letter was addressed to a microscope manufacturer named Reeves—and thus there are several interesting conclusions to be drawn.

First it is clear that Reeves must have been functioning as a maker of microscopes for some time prior to 1662 (or there would have been no standard models from which Dr Powers could have gained the working experience he reveals by writing the letter);

Second, we may deduce that the instruments (since they had a field-lens) were compound microscopes of a fairly sophisticated nature;

Third, we may see how these facts are accepted without any due comment from Powers—it is not as though Reeves was being consulted as a famous pioneer—which suggests that microscopes and their makers were not extraordinary even, say, in 1660 or perhaps earlier.

This postulated picture of the state of the art tends to give currency to the microscope significantly earlier than the dates that are widely accepted. And in its own way it provides further confirmation that microscopes 'just growed', rather than being invented through the specific genius of any individual. Noblemen were amongst those to be presented with microscopes as objects of interest, and the diarist Pepys had one. So, from the time when the very first mention of the word 'microscopia' appeared in the literature (in a letter written by Giovanni Fabri dated 13th April 1625), to the time when the instrument was becoming accepted—albeit only as a toy, almost—was only some thirty-five years. But by this time, the writings of those who had made microscopical observations and the degree of interest that the professional scientists were showing had, clearly, increased; men's minds were now attuned to the possibilities of microscopical investigations, and their ways of thought, tempted along by the occasional references to microscopic creatures and fine structures that must by then have occurred in scientific conversation, had paved the way for the next moves in the game.

Robert Hooke was born in 1635. This was in an era of great change for England. Only ten years before, James I had been on the throne—almost the last monarch to adhere to the 'divine right' of kings.

But changes of a fundamental kind were afoot. There were demands for democracy from the people, calls for greater conceptual freedom amongst the learned; and everywhere there were warring factions: Catholic against Protestant, Puritan against Nonconformist, monarchist against democrat . . . the choice was limitless. At the time of Hooke's birth, there were no sessions of Parliament at all, as the young King Charles I tried to reassert his power over the people. Matters came to a head at Edgehill on 23rd October 1642 with the first battle of the Civil War, which culminated early in 1649 with the execution of the king. Shortly thereafter Parliament voted that the House of Lords was 'useless, dangerous and ought to be abolished'.

The new tide of liberalization in thought had left its mark on scientists, and a group of men formed an association nicknamed 'the Invisible College' in 1645; meetings were held either in private lodgings or in Gresham College, and a whole range of discussions was organized. The principle was—in modern terminology—to provide an interdisciplinary forum for the exchange of ideas, and many of the founders of the 'college' were famous names indeed. Sir William Petty was a brilliant scientific mind; Seth Ward was an astronomer of note; Dr Goddard, another central member of the group, was Oliver Cromwell's personal physician, and Dr Wilkins, a Presbyterian minister, was destined to become Cromwell's brother-in-law (a fact that he used to advantage during the years of Cromwell's power, but none the less he still managed to become Bishop of Chester after the Restoration). The leading mathematical brain in the group was Wallis, who kept his Royalist tendencies under cover until the time was more propitious to reveal them. And there were two other leading members, both of whom have a significant part to play in the unravelling of the microscope's early history. They were Christopher Wren, the architect, and Robert Boyle. These men need no introduction to the reader, though there are certainly many aspects of both characters that are not so popularly known. Wren and Boyle were close friends, not only through the 'Invisible College', and it is both interesting and important to mention that Wren, at about 1660, had a microscope of his own. Some of the drawings made with it he presented to Charles II some time later. He often discussed the subject of microscopy, it seems, with Boyle—who, incidentally, never attended any university, nor was he ever attached to any college; he was granted an honorary degree of Doctor of Physics at Oxford in 1665—and it was through this means that the subject of microscopy was brought to the attention of Hooke. In 1658, when Boyle was based at Oxford, he was approached by the chorister from Christ Church who was both interested in scientific experiments and already so impecunious as to be willing to offer his services for next to nothing. Boyle was working on air-pumps at the time and was looking for an apprentice; the outcome was obvious—Hooke became Boyle's assistant. It was a prophetic association.

In 1660 Charles II gave approval to the formation of the Royal Society. Boyle was one of the founder members, and Hooke, of course, went to London with him at this time and assisted in many of his experiments. Hooke thus watched Boyle experimenting on birds and mice which he placed in glass vessels from which he then evacuated the air; and in turn Hooke studied Wren's physiological experiments which led to the principle of intravenous injection. It is, in this connection, unfortunate to reflect that Wren is remembered by most people as an architect: he was a capable physiologist and a perceptive physicist too. Later Hooke carried out some physiological experiments himself; he punctured the lungs of experimental animals opened on the table and forced air through the trachea with one of Boyle's pumps, observing that the oxygenation of the blood continued (though he did not, of course, describe it in those terms) and the heart remained beating—in contradiction to the belief that it was the breathing movements that were the necessary stimulus. And all the time, when he was not working at these experiments, he was building himself a microscope. He ground his lenses carefully (designing a most ingenious machine for the purpose) and made the body of the instrument from cardboard. It was covered with decorated material and mounted on a base of polished hardwood. Hooke actually made several microscopes during his career, but this first prototype was enough to fire his interest. Through Boyle's influence he had been made Curator

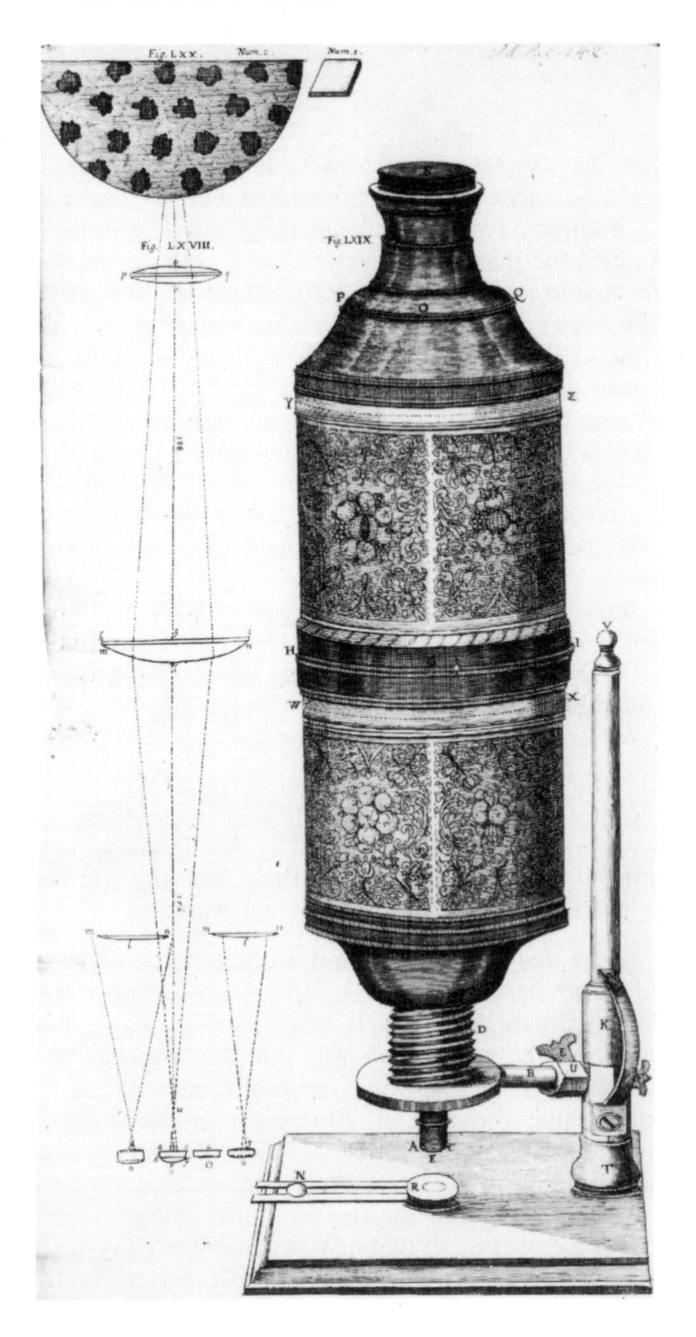

Hooke's microscope drawn by Sturm. Note the elaborate ornamentation on the cardboard body tube. From *Collegium Curiosum* (1676).

of Experiments to the Royal Society, and was so delighted with his new instrument that he demonstrated it on one of the experimental evenings. There was considerable interest aroused by the event since the microscope was still a novelty to most of the members, and on 25th March 1663 Hooke was officially given the task of carrying forward his investigations in order to publish them. Since he was probably the first professional microscopist ever, new discoveries were readily available. For example, he found that cork 'had a very little solid substance', and noted that it was composed of a great many pores. 'These pores, or cells, were not very deep,' he went on—and in this way the term 'cell' entered scientific literature. It was a matter of centuries, however, before the true significance of these 'cells' was appreciated.

Hooke's published volume *Micrographia*, which appeared under the Royal Society imprint in 1665, throws an interesting light on his motives for commencing these investigations. His introduction to the microscope through the interest of Christopher Wren is referred to in the preface of the book when he writes:

> At last, having been assured both by Dr Wilkins and Dr Wren himself, that he had given over his intentions of prosecuting it, and not finding that there was any else design'd the pursuing of it, I set upon this undertaking.

It is clear that Hooke, having in his hands a uniquely simple key to the whole world of microscopic structure and at the same time a rare situation in which to work, needed an assurance that he was the only man known who could be expected to commence the great work of discovery and documentation that lay in wait. He knew then that he had an unprecedented opportunity, not only to use the instrument as an object of curiosity, but to set about examining, drawing and describing microscopic structures in a systematic fashion for the very first time in history. He worked at a great rate so that he was assured of precedence, and his writings effervesce with the exuberance that a scientist may only have at such a time. He was (as he knew full well) on the verge of an important journey of discovery. It was he who, for the first time, would be able to look closely into the hidden universe.

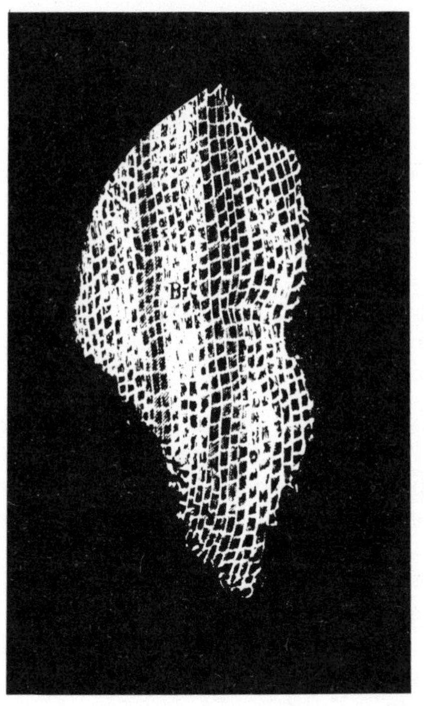

Left. Robert Hooke's drawing of a hand-cut cork section, made in April 1663 at the Royal Society. *Below*. A similar hand-cut specimen photographed by the author using a low-power objective similar to that employed by Hooke.

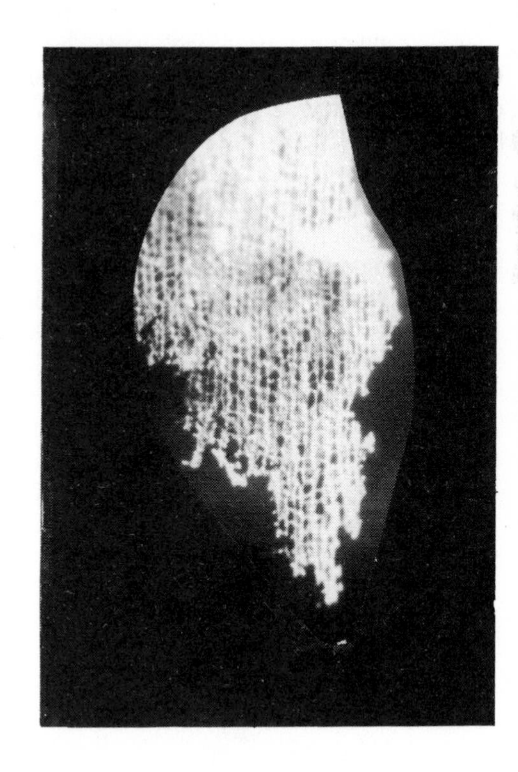

2 The Plant takes Root

MANY are the factors that sow seeds of inventiveness in the minds of men, but one above all is a prerequisite of discovery by research. And that is, simply, work. The adage that such steps are taken by the aid of 'one per cent inspiration; ninety-nine per cent perspiration' is truly the secret of progress, and Hooke—as he began his uncharted and largely random wanderings through the microscopic universe—was certainly a worker.

Spurred on by the urge to break new ground, confident that there were in all probability no rival contenders, encouraged by the work of such peerless men as Wren and Boyle, and given the tangible stimulus of a Royal Society order to produce one new demonstration each week, he began to make startling and new observations at an almost unprecedented rate. His first demonstration, arranged for the members to see on 8th April 1663, was typical of those specimens marvelled at by users of lenses for decades—a moss. His delicate drawing of the specimen (which I would identify as *Tortula muralis*, the common wall screw moss) is mainly a study of the features that are visible to the naked eye—but with one interesting exception. Though the leaflets are in the main drawn to look like conventional lily leaves, or something similar, one of the figures shows them at a higher magnification and five of the little leaflets are drawn in greater detail. Their entire surface is shown to be composed of small rectangular structures, on which Hooke does not comment in his accompanying description. Though he did not know it, he had already made an important discovery.

He was impressed by the smallness and neatness of these tiny box-like structures, and in the next few days he turned his attention to another vegetable specimen, and looked for the 'little boxes' again. He wrote in his notes at the time as follows:

> I took a good clear piece of Cork, and with a Pen-knife sharpen'd as keen as a Razor, I cut a piece of it off . . . then examining it very diligently with a *Microscope*, I could perceive it to appear a little porous . . . these pores, or cells, were not very deep, but consisted of a great many little Boxes, separated out of one continued long pore, by certain *Diaphragms.*

By analogizing these box-like structures to small rooms—cells—Hooke coined one of the most important and fundamental terms in the history of biological science. Yet it would be wrong to accredit him with too much in this respect: although he had observed cells and given them the name which they bear to this day, it was to be a matter of centuries before there was any real appreciation of their importance. The 'cell theory' was as yet not even a dot on the horizon.

So in the first two recorded observations of Hooke's series, he established the bridgehead between magnifiers and microscopes; between the studies of moss (which showed familiar details, but larger) and those of cork with their revelations of microscopic structure—i.e., details that the eye of man could not hope to see unaided. He carried the same concept of a porous structure into the world of mineral specimens, too, by examining a specimen of kettering-stone which he found to be 'exceeding porous'. Was it not possible, he argued, that porosity was in some way a common characteristic of many substances, living or non-living? And was it not possible that light was transmitted by the fluid in the interstices? His thoughts emerge from his writings as speculative, meandering chains of ideas; as we know in some ways Hooke was not so far from the truth. When the Fellows arrived at the Royal Society on the evening of 13th April they were shown the specimens of cork and the slides of kettering-stone alongside each other. The conversation, though unrecorded, would surely have been fascinating.

A week later he set up demonstrations of vinegar 'eels' (in reality simple nematode worms—*eel-worms* as they are called to this day) which he imagined to be a kind of miniature snake with gills. Yet we must realize that in this work Hooke was not breaking entirely new ground, rather he was following the footsteps of his professional precedents.

The moss studies were familiar to all, Kircher (as we have mentioned earlier) has described the existence of 'pores' in cork roughly eight years before, and Henry Power had written good descriptions of vinegar nematodes in the preceding decades. With the observations of moulds which he presented (with the eel-worms) on 22nd April 1663 he was striking out on his own. The descriptions are not quite as accurate as they should have been, however, and he missed many salient details that any competent observer would have ordinarily noticed at once. I fancy that his drawings show *Rhizopus*, the pin-mould; they may show *Mucor* which is a close relative, but his descriptions of the colours are inaccurate and, surprisingly, he did not notice any spores although he drew fifteen sporangia in various stages of development and maturation. His conclusions, furthermore, are legendarily quaint. 'Mould and Mushromes require no seminal* property,' he wrote, adding that moulds are produced from any kind of putrifying animal or vegetable substance. He adds, 'Having considered several kinds of them, I could never find any thing in them that I could with any probability ghess (*sic*) to be the seed of it, so that it does not as yet appear (that I know of) that Mushromes may be generated from a seed'. Had his observations been even a little more careful (it is notoriously difficult to observe any of these fungi without the mature specimens being almost smothered in spores) he would have seen the 'seeds' he was looking for—and he might have thought twice about the spontaneous generation of 'Moulds and Mushromes'.

The following week he set up a series of demonstrations of crystals (quartz, perhaps) in fractured flint, and here his true mettle as a great scientist emerged. He speculated on refraction, the nature of crystal configurations, and on through the interrelated arguments that surround these issues. It was not, it is true, the kind of definitive and profound analysis which results in the coining of an important new idea and in many ways some of his approaches were predictable in the light of contemporary opinion. On the same day (29th April) he also set up a demonstration of a 'hunting spider' which he described as having six eyes; he adds that, of the many sorts of spiders he had observed, some had six, others eight eyes; and 'others with fewer, some

*See *seminaria*, page 23.

with more'. Once again his accuracy of observation, though this can be no reflection on his undeniable industry and devotion, was not as great as it might well have been. The Araneida—the spiders—usually have eight eyes.

Here too we see the microscope being used to magnify familiar structures, rather than to reveal entirely new ones—but Hooke's descriptions of the spider's web threw up some microscopical discoveries. He observed a difference in diameter between radial and concentric elements of the web, and he made some observations of the droplets (of adhesive matter) with which the fine threads are covered.

In the following weeks he continued to investigate insects (it will be remembered that Hooke was certainly not creating any precedent by utilizing the microscope to this end) until 27th May when he gave an account of 'pores in petrified wood'. Noting that he had observed similar pores in rotten specimens of oak and other similar conventional tree species he adduced that petrified wood *was* wood, although he assumed that the 'largeness of the pores' (which is actually due to the structure of the fossilized species) was caused in some way by petrification. But he went on to advance a thesis which, in his characteristic fashion, had no immediate bearing on microscopy but which threw up some interesting evidence to suggest the true nature of fossils in general.

It was at that time widely held that shell-like remains in rock were the result of a 'kind of Plastick virtue inherent in the earth'. Hooke applied an interesting analogy in order to substantiate the rival view—namely, that the fossil shells were the remains of living creatures.

He drew up firstly the reasons for claiming that petrified wood *was* organic in origin by comparing its microscopical structure to that of living species. In this the microscope was an indispensable part of the investigation. He then listed characteristics of the petrified specimen which showed how different it was from fresh wood.

The criteria he listed were:

> First in *weight*, being to common water as $3\frac{3}{4}$ to 1 . . .
> Secondly, in *hardness*, being very neer as hard as Flint . . .

> Thirdly, in the *closeness* of it . . .*
> Fourthly, in its *incombustibleness* . . .
> Fifthly, in its dissolubleness . . .
> Sixthly, in its rigidness and friability . . .
> Seventhly, it seem'd also very differing from wood to the *touch, feeling* more cold than Wood usually does, and much more like other close* stones and Minerals.

By application of these criteria to the study of fossilized shells he suggested that the mere fact that they were different in nature from fresh specimens was not, in itself, enough to deny them a common origin. The cause, he felt, was more scientific: the shells were originally 'thrown' to their observed sites by the action of some flood or earthquake, and thence they were submerged in mud and clay. In the fullness of time, he imagined, these inorganic elements would take the place of the animal within and we would be left with 'onely impressions, both on the containing and contained substances'. Thus he demonstrated the feasibility of fossilization—the first scientific investigation of which we know where the microscope, by revealing hitherto unseen structures, had played an indispensable part.

Most of the remainder of his work was concerned with the portrayal of small familiar structures in larger dimensions than those ever seen before—his study of the flea is legendarily famous—and most of his revelations were in reality confirmations of previously held views (such as the nature of the composite insect eye, for example, which had been demonstrated many times before). But his drawings of the objects he studied were superlatively good, and his powers of conceptual inventiveness were considerable. Hooke was not a peerless genius. Some of his noted work (such as Hooke's Law, which states that the extension of a spring varies with the force extending it) was merely the paraphrasing of familiar observations, and a good deal of his views were superficial. He was not the kind of man to carry out his ideas to the ultimate conclusion, and he was always willing to end with suppositions and probabilities.

**Closeness* here means structural conformity—i.e., the lack of porosity of the materials concerned. Adjacent structural elements were 'close', in other words.

But this is also part of the scientific method, and his views on the nature of light (which he considered to be a form of 'transverse vibrations') certainly anticipated the electromagnetic theories that were later to emerge. His concept of gravity, perhaps representative of others, shows that Newton was by no means the discoverer of entirely new principles and Hooke's experiments on the nature of lunar craters led to the conclusion that either volcanic action or bombardment (although meteorites were unknown in his day) were the cause. Over three centuries later, when men walk on the moon as a matter of course, we are virtually none the wiser.

All this, together with much more, derived from his central preoccupation with microscopes and microscopia. It is in some ways surprising to find a man who, though pernickety and precise in the technical quality of his draughtsmanship, was content to be noticeably less than a perfectionist in other ways. The objective lenses he used for his observations, for instance, were generally no more than melted round beads of glass.

Yet is there not a very good reason for the deficiencies?

Let us examine the *Micrographia* again in the light of criteria that are arguably necessary for scientific acclaim. They throw an interesting light on the nature of Hooke's motivations, and give some new insight into the volume itself.

The title of the book is misleading for it does not contain exclusively 'microscopic observations' (as it might be literally translated). It is not as though non-microscopical facts and figures were creeping into the main argument, for the book is liberally sprinkled with lengthy sections on totally unrelated topics. Even in his lengthy preface he discusses meteorology at length; he describes a form of barometer he had devised and noted that the pressure of air is less when rain is imminent, and more when it comes from the east—'which having past over vast tracts of land is heavy with Earthy Particles'. Here was the very bedrock of today's weather-forecasting, even though we now know that air pressure varies with its temperature and (more especially) its water vapour content as the cyclones and anticyclones whirl about the atmosphere like ripples and whirlpools in a stream. It certainly had nothing to do with the 'Earthy Particles' to which Hooke alluded!

He describes this as 'his invention', but is similarly possessive

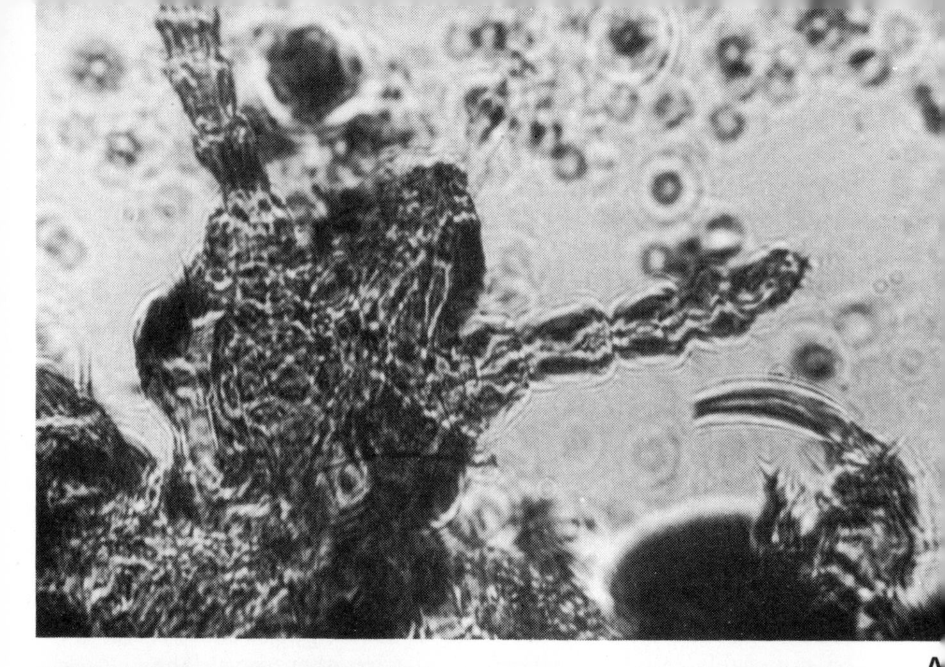

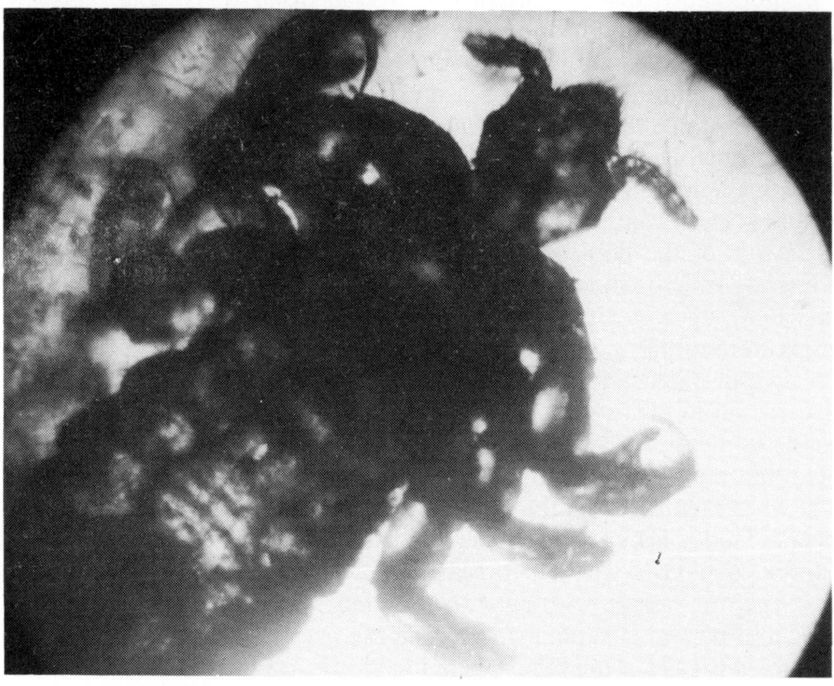

A

B

Optical microscopes and the louse.
A: Reconstructed bead-lens microscope made by the author gives a pioneer's-eye-view of the louse *Pediculus humanus*. The lens, produced by melting a fine drawn-out rod of glass but without any subsequent polishing, gives a remarkably clear view in spite of its crude nature. B: An original specimen photographed by the author using

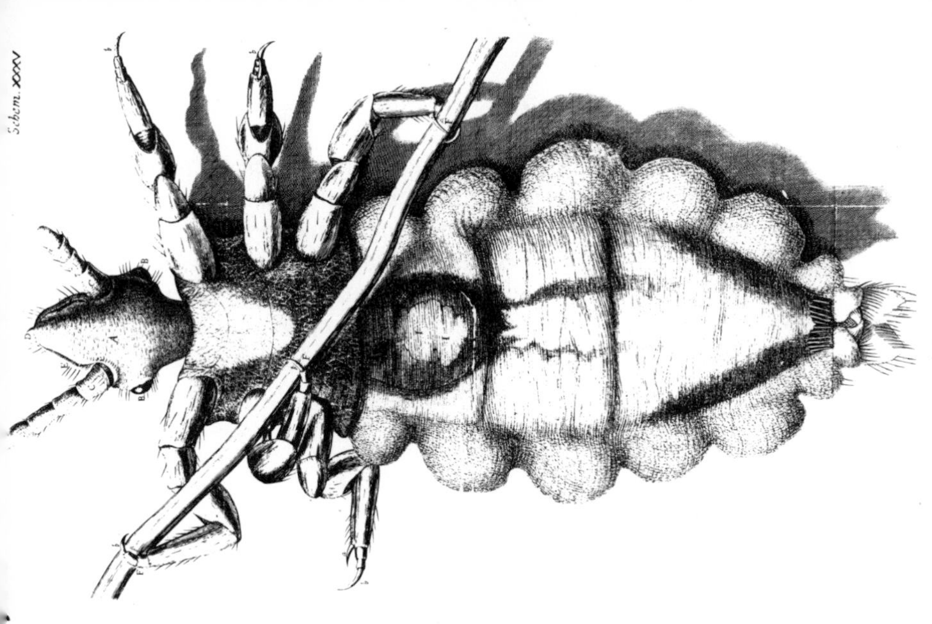

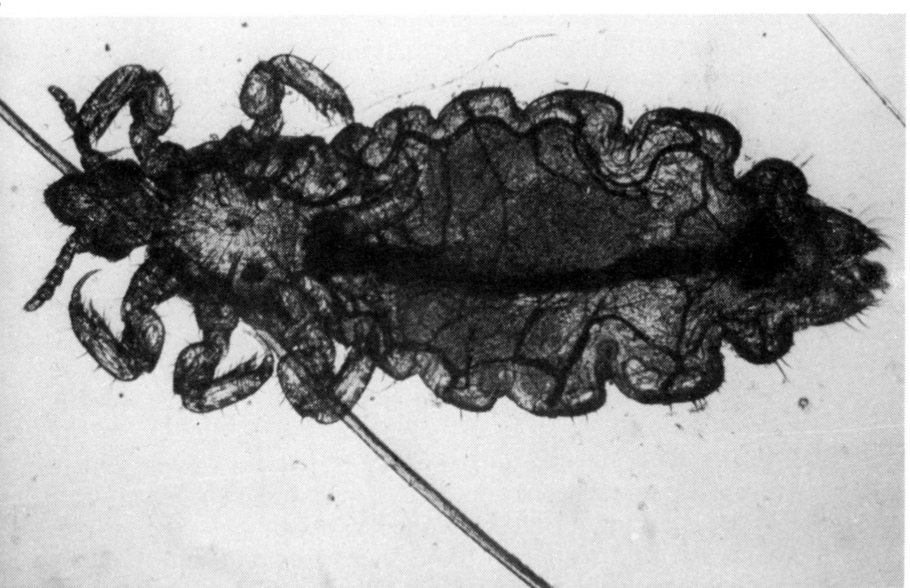

a Wilson screw-barrel microscope from the collection of Carl Routledge.
C: Hooke's fine engraving of the louse appeared in *Micrographia* (page 36).
D: For comparison, this mature louse has been specially mounted in Canada balsam and photographed using a modern research microscope.

about the use of a lamp flame, focussed through a circular jar of water, as an illuminant for his microscope. He certainly was not the first to invent this latter technique (Pliny wrote of it in A.D. 77) and it is likely that his barometer was unique only in its mode of data presentation—a small weight floating on the mercury reservoir surface moved a cord which, in turn, pivoted a needle over a circular scale. Hooke seemed determined to mention his barometer, although its link with the preface to a book on microscopic objects is tenuous to say the least.

Much the same observation applies to the rest of the book. He starts with a series of studies: 'Of the Point of a sharp small Needle', 'Of the Edge of a Razor', 'Of fine Lawn, or Linnen Cloth', and the like—familiar objects portrayed in graphically large dimensions. But the sixth observation in his book*—'Of small Glass Canes'—takes him off into a long speculative treatise on the nature of capillarity.

And so (after a mere ten pages of microscopical observations) Hooke is away at a tangent; he discusses the manufacture of lead shot, fracture in glass beads, fresh-water springs, the formation of hailstones, surface tension; for the following thirty-four pages (roughly 3½ times the amount so far devoted to 'micrographia') he digresses from the subject to a greater or lesser extent. Then, after three short pages discussing the '*fiery Sparks struck from a Flint or Steel*' he is off again through the realms of refraction, the nature of colour, both real and due to diffraction, and a range of other observations on the nature of light until, with something of a sudden jarring of the senses, we are back on page 79 with '*Figures observ'd in small Sand*' and '*Gravel in Urine*' on page 81.

Micrographia continues—with occasional digressions—through the realm of plant and animal specimens which were no doubt familiar to the reader (cork, wood, moss, the surfaces of leaves; and thence to flies, gnats, fleas and mites) ending suddenly with the change from 'Observ. LVII: *Of the* Eels *in Vinegar*' to 'Observ. LVIII. *Of a new Property in the* Air, *and several other transparent* Mediums *nam'd* Inflection . . .' of which little has been heard since. Since the phenomenon seems to be based on the refraction of rays of light through a curved path

*The first edition (1665).

(which is impossible) that is scarcely surprising. It is possible that Hooke, when writing of this '*multiplicate refraction*' as he also called it, was observing diffraction fringes, a matter which need not concern us here.*

Within the following pages he ranged from the measurement of the moon's distance from earth to the 'elasticity' of the air; to the nature of the multitudes of small stars he observed with a small telescope; and eventually to the nature of the moon itself. Tucked away amongst the legitimate microscopy are occasional isolated prognostications, at least one of which is worth quoting at length as it anticipates man-made fibres through an ingenious stroke of extrapolative logic:

> I have often thought, that probably there might be a way found out, to make an artificial glutinous composition much resembling, if not full as good, nay better, then (*sic*) that excrement, or whatever other substance it be out of which, the Silk-worm wire-draws his clew . . . This hint therefore, may, I hope, give some Ingenious inquisitive Person an occasion of making some Trials, which if successful, I have my aim, and I suppose he will have no occasion to be displeased.

That—a description of the twentieth-century artificial fibre industry—was prophecy indeed!

The book was produced with rare haste—even in today's terms. The experiments he describes were begun in early 1663, and the book appeared for sale exactly two years later, during which time the investigations were carried out, written up and illustrated, and liberally padded out with these observations ranging over the whole of science. A possible motive for these many curious facets of the book appears.

Hooke, I am sure, wanted a medium for his thoughts and ideas; he wanted to publish his own interpretation of scientific events and so to make his mark. The interest in microscopy at this time was considerable, and Hooke realized that it was this subject which would make the most propitious basis for his observations. As we have seen, he was careful to assure himself

*Those interested might care to note that refraction occurs when light passes from a medium of one density to one of another; the 'bending' of the ray takes place at the medium interface. Hooke's inflection implied a series of changes (manifested as a curve) taking place *within* the denser medium itself.

as far as he could that no-one else was likely to beat him to it; he published his work with remarkable speed in order to baulk any hidden competition; and he illustrated it lavishly with eye-catching views through the lens which would excite and interest the layman.

The views of fabrics, and particularly those of the edge of a razor and the point of a pin, were scientifically valueless—but they certainly intrigued a wide readership. So Hooke had a ready, prepared audience, an assimilable message, and he used this medium to present many of his more general scientific ideas. As it turned out his book did indeed assure him of a place in history—but one must emphasize that anyone with even a humble microscope at that time could have done the job of writing *Micrographia.* It needed time, financial backing, learned encouragement and the rest—just what Hooke had, in fact. The talents he brought to bear were basically those of industry and application, coupled with considerable abilities as a draughtsman.

Hooke had realized the importance of fashionism—of widespread popularity of an idea, of an underlying readiness to accept it—in providing new notions with an easy ride. A book of ideas would have been a poor medium indeed, alongside the guaranteed commercial vehicle of this exciting, novel picture-book, launched on a wave of fashionism.

And we must read his work with these observations in mind; particularly where his views on the physical sciences were concerned it is probably more relevant to consider them as representative of his time, rather than being wholly original in conception.

So Robert Hooke fashioned for himself a niche in science, a well-read presentation for ideas that might otherwise have remained neglected, and at the same time produced (precisely as he intended to) history's first convincing glimpse through the microscope. In *Micrographia* the instrument became popularized and the discipline itself was given impulse. Had Hooke not done it somebody else assuredly would.

Whilst the trendy scientific circles in England were studying Hooke's publication with interest—Pepys described his copy as 'a most excellent piece'—developments in Europe were following their natural, independent course. In the Royal Society's *Philosophical Transactions* for 1668 some correspondence was published which claimed that an Italian, Eustachio Divini, had discovered an animal 'lesser than any of those hitherto seen' with the aid of a microscope. Many spare-time experimenters were trying their hand at microscopic observations and the techniques of lens grinding were fast developing. Most of them, no doubt, were anonymous men working in obscurity and without purpose —and it is this element of worthwhileness which any project needs before it will be pursued with interest—and therefore with little impetus. Hooke needed the assurance of exclusivity and recognition before he started on his methodical work and many of the early amateur lens-grinders must have been regarded as cranks by their families.

But not all were. Some—like Divini—had learned affiliations; and it was through just such an acquaintance that Dr Reinier de Graaf (a leading Dutch physician) came to play an unwittingly vital role in the history of microscopic discovery.

De Graaf was a researcher of note. It was he who discovered the Graafian follicles—the bodies in the ovary within which each ovum lies—with the aid of a low-power microscope. Most of his work was done in his twenties (he died at the age of thirty-two) and he corresponded on several occasions with the Royal Society's first secretary, Henry Oldenburg. Oldenburg was a remarkable man and saw, as part of his duties with the Society, the need to correspond with as many as possible of Europe's prominent researchers. With political tensions as they were, his proclivity for international correspondence landed him in hot water on more than one occasion (and even in the Tower of London, as a prisoner of the state), but he was an honourable man and was eventually freed from restrictions. The effect of all this was to cause widespread unrest amongst intellectuals in general; both Pepys and Evelyn (the diarists) who knew Oldenburg well, wrote of the widespread 'paniq, feare and consternation' that was typical of this time.

London in particular had reason to feel highly-strung, quite apart from political issues. In June 1665 the hot, dusty summer heralded the biggest outbreak of bubonic plague that this country has seen—the Great Plague. And in the following year there was London's great fire which ravaged the centre of the city and served, incidentally, both to clear it of vermin and to raze outdated buildings ready for redevelopment.

Meanwhile Britain was at war with Holland. The Dutch had been rivals in commerce for too long, their fleet had been in the Thames when the plague broke out (so they were blamed for that) and a war almost inevitable. In the following year, 1667, they raided the docks at Chatham and burned nearly all the shipping moored there—except for one vessel, the *Royal Charles*, which they took back to Holland as a trophy.

This served to block the further progress of microscopical science, for it was in Holland that the tradition of microscopy—founded scores of years before, as we have seen—was being actively pursued. De Graaf's work was already well known in Britain's scientific circles, and in his native Holland he attracted considerable attention too. And so on 28th April 1673—when the war with the Dutch was beginning to peter out—he came to write a letter which was far-reaching in its implications. It was addressed to Oldenburg, and mentioned a 'certain most ingenious person, named Leewenhoeck', whose microscopes 'far surpass' anything seen before. De Graaf enclosed a letter from this Dutchman outlining some crude observations on moulds, bees and lice, and he entreated Oldenburg to write and encourage him in his efforts. The Fellows of the Royal Society were sympathetic to the work shown in the letter, and so Oldenburg was instructed to write back at once. This he did, and when Leeuwenhoek* had drafted out his reply he showed it to Constantijn Huygens, a statesman and writer (and father of Christiaan Huygens the astronomer) who at once wrote to Robert Hooke.

'Our honest citizen . . . Leawenhoek . . . is a person ignorant of both science and languages but of his own inclination exceedingly inquisitive and industrious,' he wrote. Within a fortnight the Fellows of the Royal Society had a chance to find out something of Leeuwenhoek first-hand, when he wrote his considered reply to Oldenburg. In his letter he explained that he had been asked to write down his observations previously, but he had never bothered to do so. The reasons he gave were, firstly, that he had no knowledge of formal language through which to

*Leeuwenhoek, like Shakespeare, had his name spelled in many different ways.

communicate, secondly because he was a businessman, and not a scientist, and thirdly, because he did not 'gladly suffer contradiction or censure from others'. He added that his memorandum to de Graaf was the first written record he had made of his work.

It was very far from being the last; he continued to write communications to the Royal Society for a period of some fifty years. They described his microscopic observations, and are legendary for their apparently irrelevant excursions through personal matters and various forms of philosophical conjecture; truly the letters were close to informal conversation in style, and bore not the least resemblance to the terse, clinical syntax of modern scientific literature. Many of the letters were published, generally in abbreviated form, in the *Philosophical Transactions*—he wrote no formal scientific papers.

But who was he? Antony van Leeuwenhoek was born in Delft, Holland, on 24th October 1632. He was apprenticed to a draper in Amsterdam in 1648, and around 1654 he set up business on his own account in Delft as a draper and haberdashery merchant. In 1654 he married and a year later his wife, Barbara de Meij, was pregnant; altogether she bore him five children—and all but one died in infancy. The survivor, Maria, cared for her father throughout his later life and was responsible for his final correspondence with the Royal Society when the old Leeuwenhoek lay on his death-bed.

When the war between Holland and England was at its height he was appointed to the post of Council Chamberlain (a position which bore some social status but was mundane in its practical implications) and in 1664 his mother died. This was the same year in which New York was founded, by the British capture of the Dutch colony of New Amsterdam, and so it can be seen in retrospect as a significant year. In 1666 his first wife died, and in 1671 he married a younger relative of hers, Cornelia Zwalmius (or van der Swalm). She seems to have been a more learned woman, and may have helped him actively in his microscopical investigations. In 1673 Leeuwenhoek was introduced into correspondence with the Royal Society, and 1680 he was elected to Fellowship—a recognition which he seems to have valued highly, as well he might, and to have taken very seriously (he is reported to have asked a friend whether he was entitled to a position of precedence over men with medical qualifications and such). In 1716 (when he was 83) Leeuwenhoek was awarded a special medal by the University of Louvain—a famous old university town seventeen miles east of Brussels—which, according to a letter he later wrote to the University, moved him to tears. Towards the latter part of the following year he wrote a nominal 'final epistle' to the Royal Society, apologizing for his ill-health and the fact that it prevented him from continuing his work. But after a short pause his letters recommenced and until his death on 26th August 1723 he continued to send more or less regular communications covering a series of new discoveries and observations. Even as he lay on his death-bed, he is reported to have asked a scribe to take down some observations on a sample of sand (sent to Leeuwenhoek for examination by a prospector for gold); two days later he was dead.

Right to the end he was lovingly cared for by Maria, his doting daughter from the first marriage, who acted as adviser, helper and housekeeper from the time when his second wife had died early in January 1694.

But from where did Leeuwenhoek's interest in microscopy spring? It certainly was not from any general movement of the discipline in his own home town, for there was none. He frequently mentioned to others that there were no other microscopists within his easy reach, though it would be wrong to conclude that this was necessarily a disadvantage. It is popular to imagine that academic intercourse is almost a prerequisite for original research, but in some respects isolation can assist as much; Leeuwenhoek, by working alone, was able to be entirely original in his approach to new problems. The lack of formal learning which is evident in his writings was a disadvantage, certainly, but how much might he have gained through his freedom from the conditioning of orthodoxy?

Leeuwenhoek mentions in one of his letters that he knew Swammerdam (who carried out some microscopical developments himself—and quarrelled with others over priority on more than one occasion) and this fact has been expanded by at least one writer into a theory that this was his introduction to the subject. But Leeuwenhoek was a good deal older than Swammerdam, and he was only in Amsterdam—Swammerdam's home town—during the latter's early teens. So any inspiration would seem to be very unlikely, or if it happened in that way the amount of knowledge Leeuwenhoek might have gained would not have been great. On the other hand there isn't any reason why the two should not have discussed the subject; my own childhood attempts to carry out investigations were very active at fifteen and sixteen, at which age Swammerdam knew Leeuwenhoek personally. Hooke mentioned on one occasion that he felt that Leeuwenhoek had been inspired by *Micrographia*, and there is certainly some evidence for this. Leeuwenhoek was in England for a short visit during 1668, and there is no evidence that he was active in microscopy before that time. He went to London when interest in *Micrographia* was at its height, and I have noticed that, once or twice, he gives explanations for phenomena which tally closely with Hooke's (though without mentioning the fact as a rule). Coincidence, chance or merely current opinion? We shall never be sure—but I feel that the implications are clear.

Be that as it may, Leeuwenhoek made a vast number of perceptive and accurate observations. He seems to have begun to take an interest in microscopy for the first time about 1670 or perhaps even later; his first letter (to de Graaf) dated from 1673, and the crudity of the observations cannot support the popular contention that he had been active for a prolonged 'silent period' before. His important work, however, lay not in the observations of insects and moulds but in the fact that he was the first person to make prolonged observations of microscopic living organisms. He was, of course, by no means the first to observe micro-organisms *per se*, but he showed a rare consciousness of continuity in his work. In the late summer of 1674 he took some samples of water from the 'Berkelse Mere', which he described as being two hours from his town. The next day he examined them with one of his lenses and observed 'green streaks, spirally coiled like a serpent', a clear reference to *Spirogyra*; he saw 'little animalcules' as well, which sound like rotifers, and ciliate organisms. These are single-celled animals which propel themselves by a covering of fine beating hairs (*cilia*, hence the name) and are generally visible to the naked eye—they can certainly be studied with a conventional hand-lens, a point which other commentators seem to overlook.

It was in the famous letter of 9th October 1676 that he first excelled himself as a discoverer of a whole new universe beneath the lens. On that date he sent a long letter, nearly eighteen pages of it, which discusses a whole range of microscopic forms of life including, almost certainly, the first reliable studies of bacteria. His account began with clear descriptions of *Vorticella*, a remarkable little protozoan which lives attached to some submerged surface by means of a spiral stalk. Leeuwenhoek's specimens were detached, and so he imagined the stalk to be a kind of coiled tail—a sensible enough inference at that time. He continues to describe another ciliate which does not seem to have been identified by any earlier writers on his work. But I have repeated his observations (which included the drying up and distintegration of the organisms) and I think there can be no doubt that he was observing *Paramœcium*, familiar to school biology students. Later paragraphs described larger organisms that also 'burst asunder whenever the water chanced to run off them'. They have been identified by Dobell as being rotifers, but they do not behave like that; however there are several very large protozoans—such as *Stentor*, shaped like a trumpet and clearly visible to the naked eye—which I have observed to behave just as Leeuwenhoek described.

Later he observed bacteria. His size comparisons with other—recognizable—organisms make this clear. But there is one interesting aspect of his work during this period which has not been examined before. Leeuwenhoek studied grains of spices and pepper suspended in water: this was probably in order to see whether (as popular teaching had it) the hot taste of pepper grains was due to sharp points on their surfaces. As these infusions of his were left standing he observed large numbers of ciliate organisms beginning to appear. Dobell, and most of the other writers on Leeuwenhoek's work, have overlooked one important fact when discussing this research—it is that ciliates feed on bacteria. The reason why they are so often found in these infusions is because of the decay brought about by the growth of bacteria; they proliferate and so act as a source of proteinaceous food for the far larger ciliates themselves. If Leeuwenhoek was observing ciliates in a state of rapid growth, it must have been in water containing large numbers of bacteria. Yet he didn't even observe them at this stage, as far as I have been able to ascertain. Why? They were certainly there.

We shall never find the answer to this and many other questions surrounding Leeuwenhoek's work. Other riddles we can,

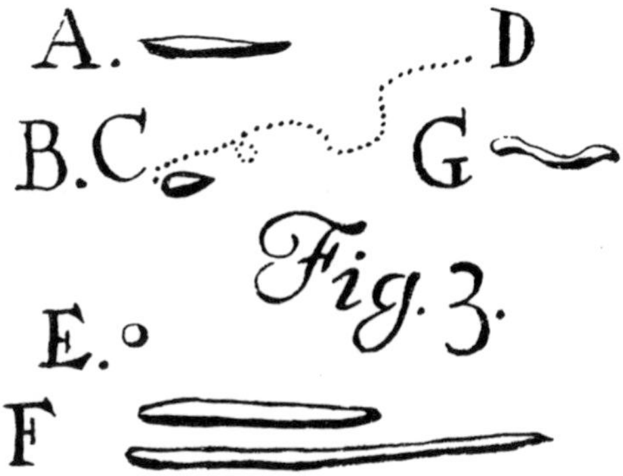

Leeuwenhoek's draughtsman made these woodcut impressions of his rough drawings of bacteria. Note the actively swimming form, indicated by a broken line.

however, attempt to solve—and some of them are of vital importance in understanding the man and his work. Of chief importance are the strange facts surrounding his means of making his observations. He never, at any time, revealed to others how he illuminated his specimens; and he often wrote of his intention to keep his 'method of observing' the 'tiniest creatures' strictly to himself. It was, in fact, a secret that he kept to his death, and which died with him; if Maria knew she certainly did not breathe a recorded word of it to anyone else.

How did he carry out his means of illuminating the specimens under observation? There seems little doubt that, at least for some of his material, he used dark-ground microscopy. This is an interesting technique, the results of which are plain from the photograph (page 112). In orthodox illumination of a specimen, the field of view is uniformly covered by a beam of light and the specimen is seen, as it were, in silhouette. In dark-ground microscopy, the light enters from the side so that the specimen itself reflects light against a black background. The result can be spectacular. Objects that are otherwise hard to see at all will show up like stars against the clear night sky of summer; indeed in recent years I have used the same method to demonstrate very fine fibrils (finer than any that had been observed previously) in coagulating blood. In normal microscopy they are simply invisible, virtually too fine to interrupt the illuminating beam. But they may still be picked out by dark-ground microscopy as fine but unmistakable images against a dark velvety background. Indeed it must be said that this form of illumination confers a great deal of aesthetic beauty to the microscopist's field of view.

Leeuwenhoek, it can be said with certainty, must have used dark-ground illumination to perceive bacteria with the clarity with which he did. Nowadays they are stained, so that they show up with ease; Leeuwenhoek could not have done this and without using such a method of illumination I doubt whether he would ever have seen them. Secondly there is a reference in his work to small particles looking like grains on black silk—in other words, having a dark-ground appearance. That technique must have helped enormously in itself.

But what of his very high magnifications? Many have observed in the past that his descriptions are so good that they leave no room for doubt that he observed bacteria—but how? His lenses, it is true, were ground with precision and were certainly the best that are on record during that entire era. But observing such tenuous organisms with a single lens is difficult, taxing and virtually impossible.

Modern microscopes are compound, that is to say they have more than one lens; the lens adjacent to the object is known as the objective, and the 'viewing lens' is the eyepiece. Compound microscopes were being made, as we have seen, before Leeuwenhoek's activities were recorded and the microscopes made during the 1660s (and thereafter) by Eustachio Divini in Italy were compound instruments. Hooke's microscope was compound, of course, although he often removed the eyepiece lens if he wished to see fine detail—neither lens was of very high quality, and the imperfections were magnified as much as the object.

But this limitation would have been dramatically less with Leeuwenhoek's precision ground lenses. As it is, with maximum magnifications of around two hundred times, it is difficult to see how this devoted Dutchman could have observed all he did: his lenses were not colour-corrected multiples of the sort we have today (and which we shall discuss later in this book). And so Leeuwenhoek died leaving a very considerable mystery behind him. It is a mystery which even his good eyesight and the use of dark-ground illumination cannot dispel—and which remains as insoluble today as it was in Leeuwenhoek's lifetime.

Or is it? There is, I am certain, an answer after all.

The mistake which occurred (which occurs still in writing on Leeuwenhoek) was to assume that he must always have used

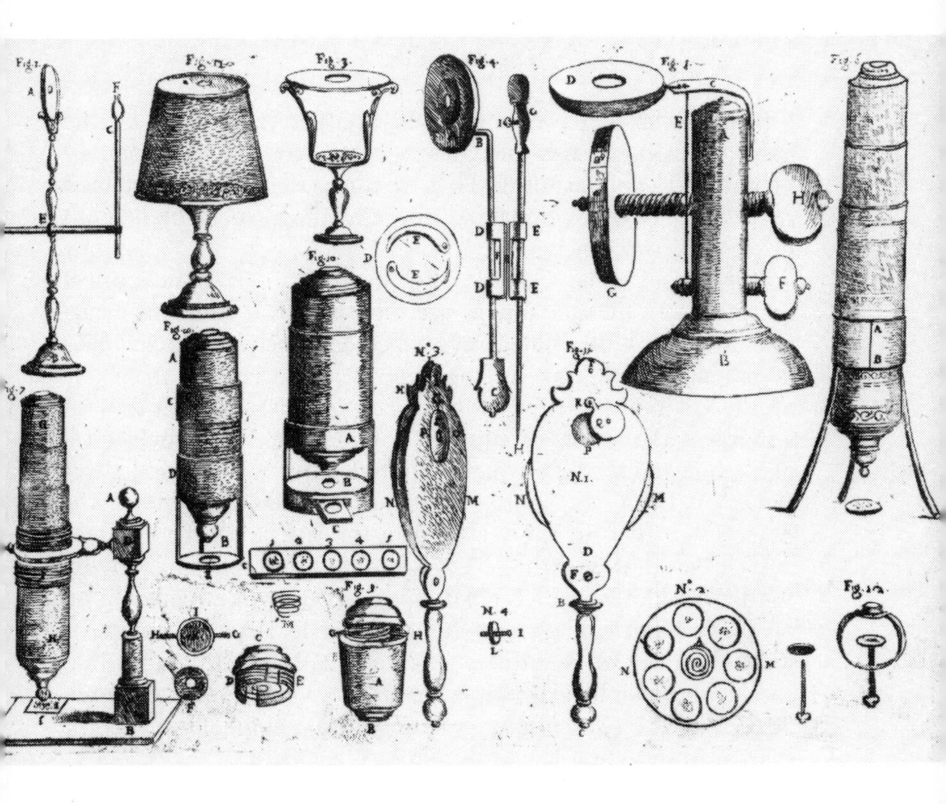

Early Italian microscopes featured some remarkable designs. Note the Divini-type microscope *(top right)*; the crude focussing mechanism and 'wheel' of specimens adjacent to it; the compass microscope *(top centre)*; and the disc of specimen chambers *(below right)*. Compare the designs in the centre with those in the illustration on page 86. Plate from *Nuove Inventioni di Tubi Ottici* (1686).

single lenses. Let us first look at the design of his microscopes: they were ludicrously put together if it was intended to carry out any focussing during the use of the instrument, since the main focussing control was situated away from the observer and was generally so close to the lens as to make it more than likely that the fingers would have interfered with the field of view—a very important consideration in high-power work. No, they are designed as though they were set up with the specimen *in situ*, focussed ready for observations, and then used with one hand. Why so? And why was the device not made to stand on its own?

I believe that Leeuwenhoek used another lens—held by the hand in front of the microscope plate—to give a secondary magnification of the image. Thus we can explain why he obtained such large magnifications and, more to the point, why he kept his secret from others.

But would it have worked? Undeniably so; basic optical theory shows that the idea is perfectly practical and I have carried out trials with single lenses which demonstrate its feasibility in practice. By taking the high-power lens assembly from one of my own microscopes I have been able to assemble it in apparatus that holds it adjacent to a mounted specimen on a glass slide and then if one holds another lens (of somewhat lower power) it is an easy matter to focus and view the object. The object/objective assembly is held in the left hand, the eyepiece lens in the right, and the object is brought into focus by moving the left hand nearer or further away—exactly like focussing an ordinary hand-held magnifying glass. The exact technique is more than we can describe adequately here, but the results that can be obtained by this simple means are staggering. The magnification of the two lenses is, one might say, summated; and the ease of observation leaves nothing to be desired.

And so Leeuwenhoek's apparently simple microscopes might have been used as compound lens systems in some of his work. We must now consider whether there is any evidence that he may have done so.

At the outset we must realize that it would have been inordinately difficult, if not quite impossible, for him to see what he did with single lenses. He could certainly have glimpsed bacteria, but some of the structures could not have been seen at these magnifications. The use of a single lens, on the face of it, seems unlikely in the light of his results.

Secondly we have Leeuwenhoek's word for it that there was a mystery technique that he would not divulge.

It was more than merely the means of illumination, too, for he wrote on one occasion that his *method* of observing the smallest animalcules he kept 'for himself alone'. Visitors found the same reluctance, for though he often showed them—with pride and with no hesitation—larger specimens such as insects' heads, plant and animal fragments and the like, which could be observed satisfactorily with a single lens (and which were sufficiently bulky to make a change of focus unnecessary during observations by visitors) he never showed to anyone the smaller specimens, nor would he divulge the means of observation. When the German Zacharias von Uffenbach went to see Leeuwenhoek in December 1710, he was shown maggots, crystals of gold, the scale of a fish, epithelial cells detached by scraping the skin, the eye of a fly and the like. Nothing, in other words, that needed a high magnification (some of which *cannot* be examined satisfactorily with anything much above, say, twenty to fifty times). Indeed it is worth pointing out that clarity, more than mere magnification, is important in microscopy; and a great many specimens (known as macroscopic, rather than microscopic) necessitate a low magnification for their proper inspection as the shallow depth of focus and limited field of view automatically introduced by higher magnifications destroy the general view. Many minerals, surfaces (such as human skin), crystals (such as snowflakes) and the like come into this category.

In 1685 Leeuwenhoek showed some of his microscopes to Constantijn Huygens (brother, not father, to Christiaan the astronomer) who later wrote that, after showing him two or three, he took them away and went off to fetch a few more, adding that he didn't trust people (especially Germans) and feared they would make off with them or become too curious.

Two brothers, Thomas and William Molyneux, went to see Leeuwenhoek in 1685. Thomas wrote afterwards some significant words which seem to confirm Leeuwenhoek's desire to keep secret some important development, for after a description of the standard Leeuwenhoek microscope he added: 'These are they which he shows to the curious who come to visit him, but besides these he told me he had another sort which no man ever looked through, setting aside himself . . . he assured me they performed far beyond any he had shown me.' And William stated, 'But for his best sort [of microscope] he begged our excuse in concealing them.'

So it would seem unarguably clear that the old man was making no secret of the fact that something, whatever it might be, was going to be kept up his sleeve. The mystery is real, therefore, and not imagined. But what evidence is there that Leeuwenhoek ever experimented with two lenses, instead of the single one, as in the accepted view? There is not a great deal—but there is some. In 1850 Hartig wrote an account of the catalogue of Leeuwenhoek's microscopes, drawn up for the auction which followed his death. In his essay these words appear: 'Two microscopes are specified as having two glasses, another three. It thus appears that Leeuwenhoek also manufactured doublets and triplets. . . .' And we also have an intriguing comment in Jean Cornand de la Crose (writing in a short-lived journal entitled *Memoirs for the Ingenious*) who in an otherwise rather disjointed account writes of Leeuwenhoek's having 'a glass or two', and a similar comment in Uffenbach's account. 'He also had some microscopes with double glasses, which were double, and the lenses were separated inside by their proper distance,' he wrote. There is therefore evidence that Leeuwenhoek *did* use systems more complex than the single lens.

The most important point of all remains—would it have worked? The main limiting factor that pioneer microscopists came up against was the sheer deficiencies of their lenses. Most of the lenses used at that time were, as we have seen, no more than melted beads of glass. I have made some experimentally myself and they do have quite startling powers of direct magnification; but when used in an array of lenses their deficiencies become so manifest as to preclude the opportunity of making any clear observations. Any deficiencies in the objective lens will be greatly magnified by the eyepiece, so that any gain from extra magnification is more than lost through the loss in image quality. The main detraction is due to spherical aberration, which is a consequence of using round lenses. The rays refracted from the outer region of the lens come to a focus in a different plane from the light rays that have passed through the centre. The appearance is demonstrated in the figure (page 71), which shows a reconstructed picture of the kind of view that Hooke must have had. The way in which this is overcome is to grind the lens profile, using fine abrasive powders of the kind found in chromium polish, until the configuration is such as to bring the rays more nearly in focus at the same place. The improvement in the picture quality is considerable.

Robert Hooke, and many other workers at that time, was too impatient to spend time grinding lenses to these fine limits of accuracy, or perhaps they felt that their simple glass beads were good enough; but Leeuwenhoek was dedicated to his task and he spent much time in perfecting the shape of his lenses so that they would give the best results possible.*

Contemporary records of Leeuwenhoek's lenses testify to their quality. Martin Folkes, who wrote an account of the microscopes presented to the Royal Society after Leeuwenhoek's death, stated that the 'Glasses are all exceedingly clear, and show the Object very bright and distinct, which must be owing to the great Care this Gentleman took, in the Choise of his Glass and his Exactness in giving it the true Figure'. He adds later that the magnification of the lenses was not as high as for many melted beads, 'yet on the other hand, the Distinctness of these very much exceeds what I have met with in the Glasses of that sort'. There are several other accounts too which confirm the great visual clarity of these lenses.

Yet their magnification was not so very high (other writers who examined his instruments discovered that the lenses magnified roughly two hundred times, some a little more; and one had the capacity to see details as small as 2 μm) so the need to use ancillary lenses must have become increasingly apparent to Leeuwenhoek as the details he wished to observe became finer and finer.

The most interesting fact which remains is that, when he died, a considerable number of lenses which were mounted between plates, but which had none of the clamps and screws with which to support the specimen, were found. These might have been used as the 'secondary lens' which I believe he employed. So there are very valid technical reasons why he could, with benefit, have used two lenses in this way; there is evidence that he did know of the possibilities, since some of his microscopes had double lenses; it is impossible to see how he could have made many of his most significant observations otherwise, and of course he reiterated to many contemporaries his intention to keep his 'best method' secret.

Did he actually do this, however? We shall, of course, never know. There is only one hint in his writings which points in the direction of confirmation, and it occurs in a letter written by

*Perhaps one ought to add that he made the metal parts of his instruments from the ore itself, which he refined to extract the metal in his own home—a rare example of devotion to the task in hand.

Leeuwenhoek to the Royal Society in 1717. He stated therein that his hands were too weak to enable him to continue his work. (He soon recovered from this phase, of course, to continue his observations for another five years.) The means of observation I postulate requires the use of two hands, and it is interesting that Leeuwenhoek does not write to say that his *hand* is weak, which would arguably have been the case if he peered up through his specimens with the single, simple microscope.

I believe that he did use two lenses, his prefocussed simple microscope—with the specimen attached—being held in his left hand, and a second lens (close to the eye) in his right; this could have given him the extra mysterious magnification that has eluded us for the past three centuries and more, and introduces an interesting point into the argument. It is this: because Leeuwenhoek's microscopes were themselves generally simple, everyone has assumed that simple they must have been. But there is no real reason at all why he could not have used compound lenses in hand-held plates; the assumption that he had no choice is basically invalid. The lenses do not necessarily have to be connected by metal, but they can be joined by man himself in the process of observation; and the crudity of Leeuwenhoek's metalwork (none of his microscopes were said to be well finished) does show that the hand-held arrangement would have saved him from all the complications involved in making a focussing microscope as such.

So there is one answer to Leeuwenhoek's mysterious method of observing. But another mystery, one which occurred when he was long since dead, remains. It is that almost all his microscopes have disappeared—and some under very suspicious circumstances. After his death a case containing twenty-six of the little instruments was sent to the Royal Society, where they were examined and guarded with care (though as curiosities rather than instruments of scientific progress) until around 1820. What happened then is a mystery; it was once alleged that Baker (who examined and reported on them in 1740) had kept them, but they were certainly in the Royal Society's collection after he had finished with them.

In 1853 an isolated (and, it seems, unsubstantiated) report by Haeser stated that the collection was then 'in the British Museum', but the final record is that of Clay who in 1932 claimed that they had been lent to 'the celebrated surgeon Sir Everard Home, but we do not know what has since become of them'. Perhaps to this day they lie, forgotten and overlooked, in a dusty London attic.

Similar obscurity surrounds the fate of the various instruments given to notables, such as the pair reputedly presented to Queen Mary II (of William and Mary fame). They too have vanished from the Royal collections.

After Leeuwenhoek's death (and two years after Maria had also died) an auction was held at which 247 microscopes and 172 lenses mounted between plates—these may have been the 'eyepiece' lenses—were sold to a total of forty-two Dutch citizens. Almost all have vanished since. One was reported to belong to the German optical firm Zeiss, another was stated to be in the Deutsches Museum in Munich, and five are apparently in Dutch collections. One other was reported to be in the collection of M. Nachet of Paris, and there are several imitations (and perhaps some forgeries too) dotted about in other private collections. But the great majority of Leeuwenhoek's lenses known to exist (approximately 450) have vanished without trace. It is, in numerical, scientific and monetary terms, a very sizeable mystery! All we have in Britain are the records of those who examined the microscopes centuries ago—and there is one final interesting factor which seems to add weight to the theory I proposed above, namely that Leeuwenhoek used compound lens systems, held in the hand, to observe very small specimens. Almost all the mounted subjects that remained on his microscopes were conventional, macroscopic structures (pores (i.e., vessels) in wood, insects' parts, minerals, etc.). There was not a slide of bacteria, protozoa or any other truly microscopic specimens amongst any of them. The secret, then, was well kept.

By the end of the 1600s, Leeuwenhoek, the mysterious isolated Dutch draper, was virtually the only microscopist of note in the world. Would he have become so without the support and encouragement that derived from his association with the Royal Society? Perhaps not. A microscopist named Mellin made, around 1680, lenses which were later stated by Nehemiah Grew at the Royal Society to be 'the best I ever saw' but nothing further has been heard of him or his instruments. And there were many others too, who worked as inspired (sometimes, perhaps, as uninspiring) amateurs—that is to say they dabbled, rather than studied—and between them the bedrock of microscopy was securely laid down. Swammerdam, a contemporary of Leeuwenhoek's, devised very many ingenious techniques of microdissection to help him with his work on insect anatomy; his microscopes were mechanically superior to Leeuwenhoek's, though not as efficient optically, and were made by Musschenbroek at Leyden. In London there were makers of microscopes who produced them wholesale, we must remember, from the 1660s onwards.

Yet by the end of the century it was generally accepted that Leeuwenhoek was the master of the discipline. In spite of the others who were also peering through lenses and the (by then) large number of microscopes in existence, he was the acknowledged leader of them all—and why? Simply because his capacity for work, his enthusiasm and his unremitting patience, coupled with the spur of interest from the Royal Society, instilled in him, above all others, the desire to succeed. His lack of any desire to emulate the academicians of his day, his refusal to ape the social or scientific ways of the learned establishment, and the consequent lack of sophistication, left him unassailed by orthodoxy and free to work alone—making discoveries, founding principles, observing what was to be observed without the strictures imposed by the conformist academics of his day. Nobility, academic distinction, paper qualifications were all beneath his dignity. He embodied great powers of observation. And, perhaps more than anything else, he was not impressed by abstract notions of 'authority' and 'status'.

'Others don't believe what I say,' he used to state, 'but I do not care: I know I'm in the right.' So this humble, hard-working man came to found a well-deserved niche for himself in the history of the microscope. He was not the first microscopist by a long chalk, and he is certainly by no means the 'founder of microscopy'. He was not the first man to observe microscopic organisms either, nor the first to write about them. But, by showing a conscious awareness of a vast, undiscovered universe and by showing a grasp of microscopic organisms in a conceptual sense, Leeuwenhoek put microbiology—the study of living micro-organisms—well and truly on its feet.

There was, then, no founder of microscopy; that arose out of the coexistence on this planet of lenses and men to look through them. Hooke, we can rightly claim, was a pioneer of the discipline (quite calculatedly, as it turned out); but Leeuwenhoek was the first to grasp what lay in store. And he, without question, was the midwife of microbiology.

Yet it is only since the 1920s that Leeuwenhoek has begun to become a figure in microscopical teaching. Now he is mentioned in children's books, encyclopaedias and popular works as 'the man who invented the microscope'—which is, of course, absurd —but in the later years of the nineteenth century he was still a little-known figure: a pioneer who had never been heard of by most scientists. It was in the 1920s that the ground-swell of discussion broke into a wave of fashionistic acceptance—and since then no word of other than the most fulsome praise seems to have been uttered on the subject.

It was Clifford Dobell who wrote the first detailed biography of the Dutch draper-microbiologist. The work is a study in hero-worship, a textbook example of 'pedestal-putting'.

Dobell was a Fellow of the Royal Society and a noted microbiologist, yet his work soon betrays a startling lack of objectivity. He damns any previous commentator who had suggested that Leeuwenhoek's methods might be less than perfect with the off-hand assertion that 'they warrant no further publicity'. He considers the suggestion that Leeuwenhoek might have had 'Jewish blood' and—in light of the wave of fashionistic anti-semitism current at that period—seems horrified to the point of theatrical partisanship at the very suggestion. His evidence to the contrary, which he states to be 'overwhelming', is superficial when we get down to it.

He describes other scientists as 'superior' or 'condescending' if they reflected anything less than hero-worship for the man. Dobell is quick to identify organisms in Leeuwenhoek's writings without too much evidence. Thus when Leeuwenhoek writes of witnessing copulation in micro-organisms and compares it with the clarity of observing that of 'flying creatures', Dobell states that he means insects. But I believe that Leeuwenhoek meant just what he said, no more, no less. He was stating that ciliate copulation is as easily seen as that in flies, beetles on the wing and birds—and there is no reason to assume that he was trying to imply specificity. He was not.

When Dobell comes across writers or commentators who wrote of Leeuwenhoek in cautious terms he becomes sardonic. Thus when one historian misspells his name Dobell writes in a footnote: 'His name is spelt 'Leuwenhoeck' by Uffenbach throughout—a mistake which I have taken the liberty of correcting wherever it comes in his narrative.' This is a particularly condescending comment, particularly when the name is spelt differently by innumerable devotees who are allowed to escape without castigation—and more so when we remember that Leeuwenhoek himself spelt his own name differently on different occasions, too!

After his visit to Leeuwenhoek (page 59) Uffenbach seemed to have mixed feelings. As a result he is heartily denigrated in Dobell's book. Yet many of Uffenbach's points are entirely valid. For example, he corrects Leeuwenhoek's view of the skin 'scales' thus: 'What the good Mr. Leeuwenhoeck takes for scales are really only the particles, or scurf, from the outermost skin . . . as it dries up and peels off under the influence of the external air . . . though it always forms anew underneath'. To add to the calumny, Dobell further takes the poor man to task for using Latin expressions (which were very much more usual three centuries back, of course) and dismisses them as 'pedantry'. Yet throughout Dobell's work appear similar expressions, *lapsus calami*, *sub voce*, *in litt.*, and so forth. A strange conflict of attitudes is revealed.

In more recent scientific documentation—and elsewhere in society, of course—we may find innumerable examples of this sort: where some figure has become revered out of all proportion to his true status. The very real problem posed by the vagaries of fashionism militates against objectivity even in the field of science. This is a disturbing fact, and one with which we have not begun to grapple. Yet it inevitably means that much of our knowledge of the development of science, the emergence of scientists, and the very nature of the scientific process is not based on rational, objective, realistic considerations; but springs from an unpredictable and childlike set of unquantified instincts.

As we shall see later in this volume, our attitudes towards many legendary figures in science—Fleming, Pasteur and the rest—owe more to the fashionism of their names than to the reality of their lives. This distorts our concept of history.

And when the same proclivity extends to research, so that antibiotics or cervical cancer smears become popular purely out of fashionism, and not for scientifically valid reasons, we are entering dangerous ground. This is not just a form of retrospective refinement of attitudes, then, but a fundamental flaw in our concept of reality.

The hundred years or so after Leeuwenhoek's death are a further example of this, in fact. The microscope fell from fashionism during this period and so—for all its potentialities—it was ignored. This, as we shall see, cost us dearly in terms of progress towards humanitarianism and the new insight in medicine.

3 The Lost Century

ROBERT HOOKE, in his old age, was content in the knowledge that he had put microscopy on a sound footing. He was never sufficiently enthusiastic to carry out any profound further developments once he had established his precedent. Leeuwenhoek, as we have seen, used some far-reaching techniques and a considerable degree of insight in launching the new science of microbiology; he was satisfied in the veracity of his observations, even though he was sure that some of them would take time before they were universally accepted. Both died, it seems, feeling that they had begun to unfold a vast new dimension, with incredible potentialities for mankind in general.

It did not happen. By the time Leeuwenhoek died he was the only active microbiologist in the world. His discipline died—for the time being—with him, for mankind was not yet ready for it all. It was going to take time for the spreading ripples to exert their effect, and the natural inertia of conservatism that has ruled our species for countless centuries was too big an obstacle. It was going to be several centuries before its true impact was realized and, as we shall see, even today in the 1970s there are residues of medieval traditionalism of a kind quite incompatible even with Leeuwenhoek's simple teaching.

Although the progress of microbiology came to an almost total halt in the years that followed, the basic technology of microscope manufacture went ahead. The methods of obtaining clearer images were gradually worked out and as technology improved and better techniques gained a wider currency, the discipline developed through a process of evolution. Animal evolution, and that of plants too, tends to hinge to a considerable extent on the concept of 'survival of the fittest'—or perhaps that should read 'survival of the best adapted', since dinosaurs were certainly 'fit'.

There is a tendency for better adapted—i.e., functionally successful—species or varieties to proliferate, whilst their less favoured contemporaries fall by the wayside. And this is directly analogous to technological evolution. Processes, innovations and techniques of all kinds find that their success in the face of impractical alternatives is the key to progress—and so in the practical field we have a system of 'survival of the fittest' once again. It is particularly interesting to compare the explosive upsurge of, for instance, the motor-car—as it became adapted for its purpose—with that of the mammals, or man himself.

As technologies adapt and progress, they give rise inevitably to new developments which in turn facilitate further advancement. The need to transport men and materials in primitive societies gave rise to the need for a wheel, which eventually made the wrist-watch possible; and, in microscopy, the need to see more clearly and the annoyance of unstable instruments made men turn their attention towards mechanical developments which, in due course, made the most far-reaching discoveries possible. Yet initially the developments were not made with this in mind at all; it is part of the leap-frog chain of reaction where men use 'necessity' as the mother of invention, and the resulting child as a stimulus to further conceptual advancement. We had lenses, which as practical entities enabled men to undertake the intellectual leap into microscopic consciousness; and that in turn stimulated the next round of practical progress. But these newly developed microscopes had to wait a century or so before the next leap took place.

In the seventeenth century several microscopists rose to prominence as manufacturers of instruments. Musschenbroek, the young contemporary of Leeuwenhoek's, made microscopes for Swammerdam which featured a single lens; his low-power instruments held the object to be examined on a flexible arm made from a series of ball and socket joints which became known as 'Musschenbroek's nuts'. Later, the object and lens were held in their correct position in the manner of a draughtsman's compass, with the object on the point of one arm and the lens on the other.

Many different manufacturers produced such instruments, which are not uncommon as collector's items today, and they were refined by means of various focussing and lens attachments.

But they were toys; novelties. As scientific instruments they were only of limited value, and no discoveries of note were ever made with them.

It was still the lens which proved to be the biggest practical problem. One ingenious amateur experimenter named Stephen Gray, at about the end of the century, was so irked by his inability to melt smooth and regular glass lenses that he attempted to use a drop of water held in a wire loop as a lens. It worked surprisingly well, and a number of different types of 'water microscope'—most of them consisting of small holes in brass plates—were manufactured.

But the next serious development that we should record was the invention of the screw-barrel microscope. Many will have heard of the 'Wilson Screw-barrel' as a classical stage of microscopical evolution, but Wilson (who was a noted manufacturer

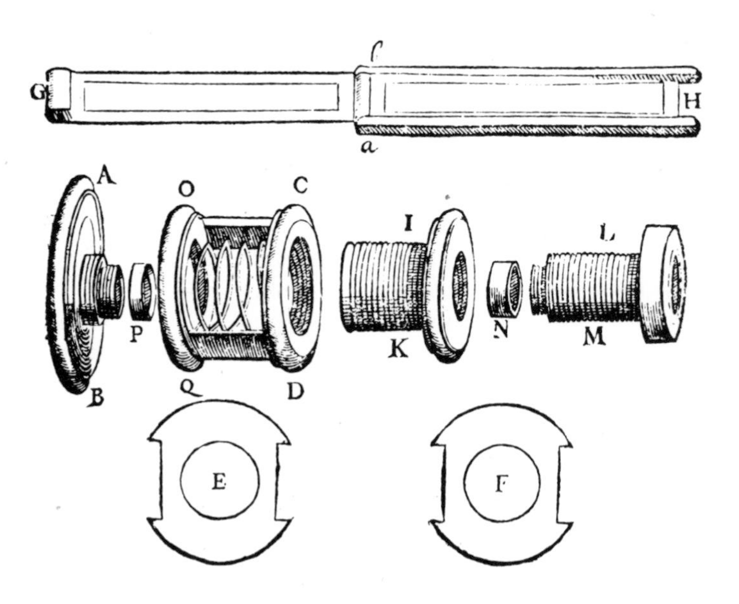

Hartsoeker's original design for a screw-barrel microscope. Note the specimen slider-holder (GH) and the plate housing the single lens, AB. From *Essai de Dioptrique* (1694).

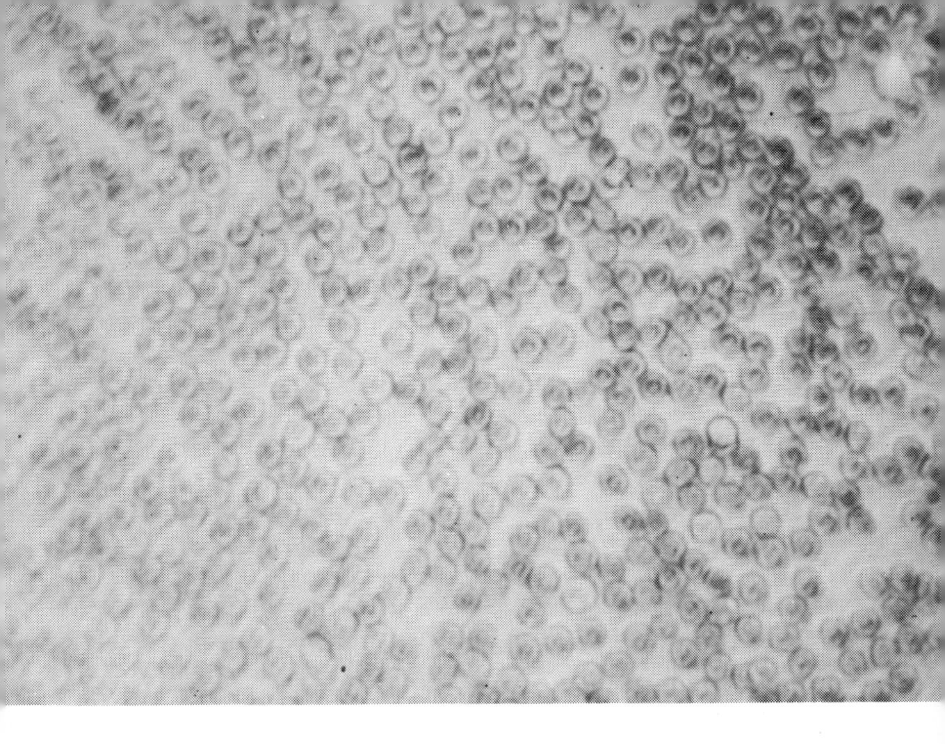

A blood-smear photographed by the author with the Wilson screw-barrel microscope (see also illustration on page 44). This is the highest power obtainable with this single-lensed instrument. Compare with the results obtained at the limits of the modern optical microscope in illustration on page 112.

of microscopes) was not the inventor of this type of microscope at all. His designs were founded squarely on those of Hartsoeker, another of Leeuwenhoek's 'rivals' and therefore denigrated heartily by Dobell's famous study. He was in fact a most able microscopist. But once again the development has its roots further back in history. It can be traced indirectly from microscopes made by Divini in Italy which were modified around 1670 by Campani to feature a screw-thread focussing arrangement. By the end of the century he and Bonanni had separately produced microscopes with screw-barrel focussing and Hartsoeker made recognizable screw-barrel microscopes in the 1690s. The origins of the term and the simple means of operation for these microscopes are both clearly evident from the diagrams (page 70).

In the *Philosophical Transactions* for 1702 appeared Wilson's first specifications for the screw-barrel arrangement, and with this announcement the first commercially successful microscopes of this kind became available.

Many manufacturers subsequently produced similar instruments, and this basic concept in design became the most widespread of all early microscopes. But those commentators who have in the past drawn attention to the apparent plagiarization of Wilson's designs are overlooking one important fact—that is, that he did not claim to be the originator in the first place. Be that as it may, his name is still the best known (even though signed instruments are very rare); and his popularization of the screw-barrel microscope made it the 'popular' design for more than a century. King George III owned one, though his wife, Queen Charlotte, preferred a compass microscope made by Cuff. Clearly—as instruments of entertainment—microscopes were ubiquitous.

Many manufacturers (such as Scarlett, Lindsay, Joblot and others) took the basic design and developed it further; the 'scroll-microscope'—in which the instrument was attached to a floral metallic scroll with the mirror at the other end—was one variety, and tripod-mountings were another. A little-known manufacturer named Brander produced a brass device which was extremely well made and advanced; it was the first of a type which today's eye would recognize as 'a microscope' in the conventional sense.

Meanwhile the more complex compound microscope was pressing ahead too; many manufacturers throughout Europe made variations on the theme illustrated by Hooke's microscope, and different only in relatively minor details. Then, in or around 1700, an instrument maker named John Marshall produced a well-made and popular 'double microscope for viewing the circulation of the blood' in the tail of a fish, and around three decades later the London manufacturer Culpeper announced a further development, in which the microscope body, the stage and the mirror were all aligned in a tripod stand; the type became known as the Culpeper microscope wherever it was manufactured, and it had copyists in almost every country in Western Europe.

The demonstration of capillary blood flow in the tail of a small fish was (and still is) a very remarkable sight to the un-

Scroll-mounted microscope designed by Cuff and published in *Description of a Pocket Microscope* (1744).

Adams's 'New Universal Single' microscope, showing stage designed to take sliders (one is seen in position) and a rotating array of simple lenses of differing powers—the forerunner of today's multiple nosepiece (illustration, page 185). From *Micrographia Illustrata* (1746).

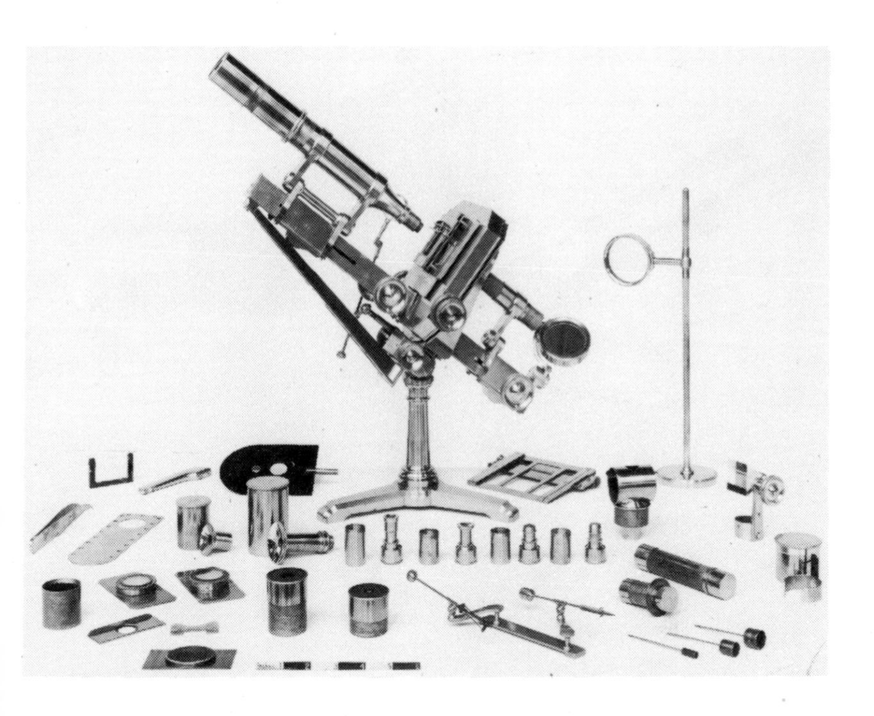

Microscope by Hugh Powell, 1840. This instrument established him as a foremost manufacturer, along with Ross and Smith. Note mechanical stage and *(lower centre)* stage forceps. The prism attachment *(right)* was used to assist in drawing specimens to scale. Science Museum Photograph.

initiated, and a standard accessory of all these early microscopes was a brass or glass holder for the specimen. Culpeper's microscopes were often beautifully tooled and decorated and his sets always featured a small plate which often featured the words: 'this Glass is to lay ye Fish on.'

Great manufacturers abounded during the eighteenth century; Martin, Scarlett, Cuff, Nairne and George Adams (who made surely history's most grotesque microscope for George III to use in his drawing room) were chief amongst them. Solar microscopes were invented too, which used the sun's rays to produce such a bright image that it could be projected on to a screen—in the manner of today's home ciné films—and the whole gamut of technical advances (including newer and better

The first form of solar microscope designed by Cuff (1744). Sunlight reflected from mirror G provided the illumination.

means of focussing, supporting specimens and changing lenses) appeared as manufacturers sought to out-do each other.

Some writers were not easily impressed by the evidence of the microscope; microscopical revelations were ridiculed by many who should have known better. And one enterprising anonymous Englishman seems to have made a fair living out of a startlingly original piece of charlatanry: he made a mock microscope in which specimens from the patient were viewed through a concealed mirror which superimposed a view of some small creature (a woodlouse or an ant, say) hidden elsewhere in the device. He produced a series of drawings of the 'animalcules' which produced conditions such as vertigo, chronic ulcers, madness and apoplexy, looking like miniature octopuses, earwigs and caterpillars. Yet, though this man (who may have been called Boyle or Boil, if some historians are to be supported) was a fraud of the highest order, he was, interestingly enough, anticipating later germ theories of disease. It was a somewhat inauspicious way of doing so.

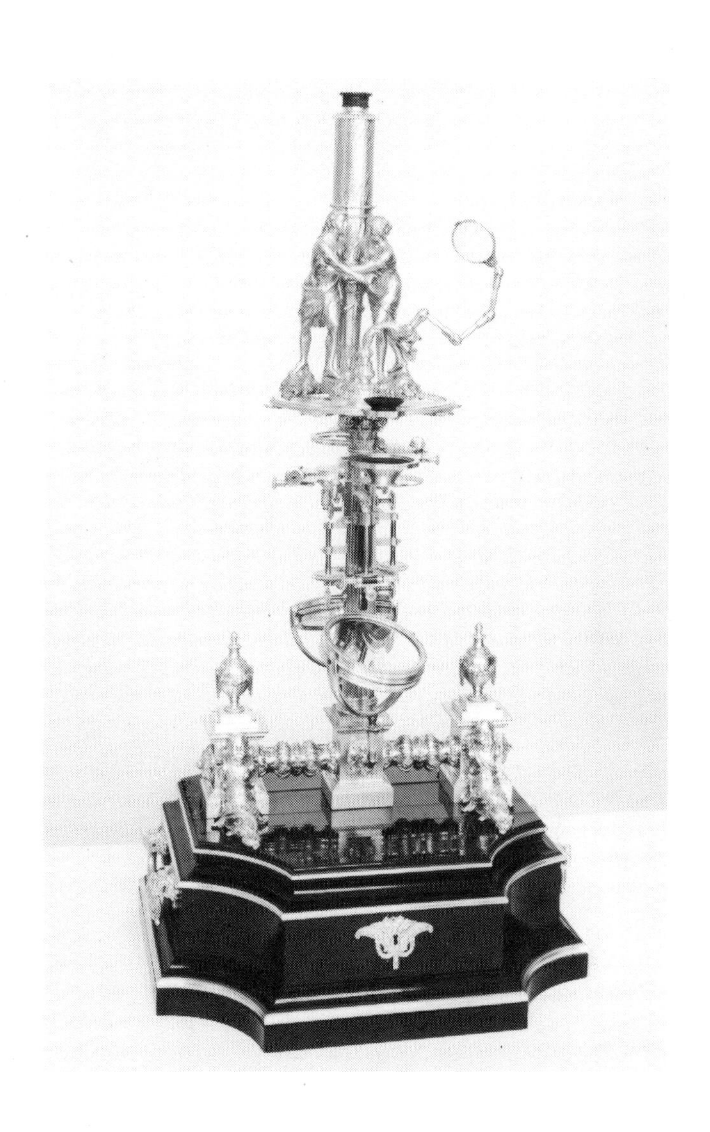

History's most ludicrous microscope—the silver instrument made specially for King George III by George Adams in 1760. Constructed out of brass and steel, covered with beaten silver, it is too tall to use if it stands on a table, yet too short if it is standing on the floor. The lenses are inconveniently placed for comfort, and the whole device is a masterpiece of impracticality. Science Museum Photograph.

This confidence trick dates from the late 1720s, and in 1757 the naturalist Nyander (a student of the great Carl von Linné, better known as Linnaeus) wrote that many illnesses, such as plague and dysentery, were caused by insects similar to the cheese-mite.

Yet there were other men who had reached these conclusions earlier. It seems almost incredible that, after Leeuwenhoek's well-known work on micro-organisms, Nyander could have failed to connect the two when he was obviously grasping at the role of germs in disease. It is unfortunate for the progress of science that the germ theory of disease took so long to emerge.

It did have several abortive attempts at raising its head. One little-known man of medicine, Dr Benjamin Marten, wrote of tuberculosis in 1720 as being of microbiological origin; he suggested that the causative organisms were 'animalcula or wonderfully minute living creatures' far smaller than could be seen by ordinary lenses. He suggested further that they could survive for a long while in dust, that their 'ova' could be transported by draughts of air, that contagion was clearly possible; and he laid down many of the fundamental doctrines of modern bacteriological theory. No doubt many others of his time had similar inclinations—it would be absurd to assume that he, simply because we know of him, was therefore the only one—but all these views were neglected, the findings were dismissed, the potentialities ignored.

At around the same time the French manufacturer Joblot published a book, *Descriptions et Usages de Plusieurs Nouveaux Microscopes*, in which he described experiments with boiled hay infusions. He noticed that micro-organisms did not arise in sealed samples kept out of contact with air in glasses sealed with parchment, and propounded that there existed 'innumerable quantities of very minute animals'. This was the first recorded experimental evidence *contra* spontaneous generation, and Marten's was the first sound bacteriological treatise on the nature of disease. Both were ignored.

Marten's view was sound, sensible, it was even topical in a sense—but it fell on unprepared ground. If any announcement does not come on the heels of some earlier work which has already begun a reorientation of attitude, then the task of gaining acceptance is infinitely greater. And so Marten's writings were soon forgotten—indeed it was not until 1911 that they were effectively heard of at all. Another anticipatory volume was that of Marcus Plenciz, published in 1762, which has been more widely discussed in recent years. He wrote that seeds of contagion and illness were spread in the air and that they were invisible, but he took the gilt from the argument, rather, by adding that much the same process gave rise to beetles, gnats and leeches. . . .

So the nature of disease was to remain beyond the pale of fashionism for a while yet. The nature of micro-organisms was a different matter, for examination of pond water, at this time a popular drawing-room pastime, was a perpetual reminder of the existence of this hidden universe. A naturalist named Redi had shown (in a series of experiments contemporary with Leeuwenhoek's observations on infusions) that decaying meat does not produce flies of its own accord, but remains free of maggots for as long as it is protected from insect attack by a sheet of muslin or a closed container. Someone at last had gone on record, therefore, as an experimenter with evidence against that of spontaneous generation. But the dogma was still widely held. We tend to imagine today that Pasteur was the man who showed the true nature of micro-organisms and where they came from, thereby destroying the myth of spontaneity. But a century earlier we may find an unsociable, perceptive Italian who tackled many of the same problems—and who anticipated many of these later developments. He was Lazzaro Spallanzani, born in 1729.

He had been determined to become a scientist from an early age, against the wishes of his father who had hoped he would become a member of the legal profession. It appears that Spallanzani was a conniving pragmatist, for he seems to have been quick to divert the whims of the establishment to his own ends. It is said that he was a fervent agnostic, yet he was able to join the Catholic priesthood for security and safety; in the words of de Kruif, he was able to 'support himself by saying masses'—and, unlike so many who had been persecuted in previous years, he could work undisturbed and largely unmolested through his facade of adherence to the Church.

Spallanzani was, without doubt, a realist. He knew that the doctrines of spontaneous generation of life had a suspiciously untenable ring about them. He knew of Redi's work. He knew enough about the dogma of the Church to find that fellow-priests were beginning to question it in many fundamental ways, and so he had a sound backing of experience from which to launch a campaign of questioning and reassessment. By all accounts his students worshipped him and on more than one occasion (when he was Professor at Reggio University, a post he took at the age of 26) his controversial views and behaviour provoked student uprisings with a strangely twentieth-century flavour to them.

At this time he was still actively building his career: from his chair in Logic and Metaphysics at Reggio he moved to the University of Modena and in 1769, at the age of forty, he became Professor of Natural History at the University of Pavia, from where he mounted animal-collecting expeditions with rare vigour. Throughout his career he continued to work, somewhat spasmodically it seems, with micro-organisms. He was sure that the strange but widespread dogma of a 'vegetative force' which made Eve spring forth from Adam, flies proliferate in cow-dung and microbes arise from putrid broth, was false; and he devoted a fair part of his life to proving it so.

His first experiments were simple. He sealed samples of meat broth and similar infusions by melting the neck of the glass containers and then he immersed them, whole, in boiling water for an hour or so. A week later he cracked open the flasks, inspected the contents under a microscope, and found that there were no living micro-organisms visible—though they invariably were present by then in similar infusions that were not so treated.

He was confident that he had found the answer to the spontaneous generation puzzle: heat a broth, so that all organisms in it are killed, and as long as it is kept out of contact with the air it will remain free of contamination.

But at the same time, in England, a truly devout Roman Catholic named J. T. Needham was seeking to prove exactly the opposite. Perhaps hearing of Spallanzani's experiments, he produced a series of specimen results before the Royal Society which, he claimed, proved the 'vegetative force' theory to be the right one. He explained how he had taken meat broth and corked it tightly into glass bottles which were then placed in 'hot ashes'—no less!—to ensure sterility. And, said Needham, many micro-organisms subsequently developed in the stored samples after treatment. Spallanzani, of course, was incensed when he heard the news and clearly could not understand why the English churchman was being so obtuse. So he devised a test of his own.

He took glass flasks and prepared infusions of meat broth, nuts, seeds and the like; and he arranged them in two groups. The first he sealed by the expedient of fusing the glass necks of the vessels so that no external air-borne entity could conceivably gain entrance, and the second series he used as a control—that is, he put them up in exactly the same manner, but this time he plugged the necks with cork stoppers as Needham had done, instead of sealing them hermetically. He was plainly convinced that it was through the minute interstices of the cork substance that Needham's cultures had become contaminated. Then all the flasks were boiled for different times, a pair of them (one sealed, one corked) for each different boiling period. And at the end of the treatment, when some had boiled for only a few minutes and others had boiled for hours on end, he cooled them gently and left them to stand.

Eventually he came to examine the infusions. He found, just as he had expected, that all of the corked samples were contaminated with a heavy growth of micro-organisms; more important still, he found that the sealed specimens that had been submitted to prolonged boiling were utterly free of any such contamination.

Perhaps most interesting of all, he observed something that startled him a good deal. In some of the sealed flasks which had been boiled for, say, less than five minutes or so there were quite heavy growths of micro-organisms—and with some surprise he realized that some species of microscopic organisms could indeed withstand the temperature of boiling water—and yet survive. The knowledge of this fact was to play a leading part in later developments in bacteriology, and it was useful to Spallanzani himself in some of his later controversies.

But the outcome of his experiments was clear, he felt; prolonged boiling would effectively kill all forms of dormant life in such a liquid, and if it was kept totally out of contact with the air it would remain free of contamination. Needham's short-sighted and biased experiments, which had ignored so many important factors, were patently foolish.

But the Englishman was not to be put off so easily. Working with the French Count of Buffon, G. L. LeClerk, who provided him with patronage and publicity, Needham came up with his answer. It was an elaboration of the 'vegetative force' theory, and soon it was on everyone's lips. Clearly, it was said, Spallanzani had boiled his infusions for so long that the vegetative force was destroyed and the power of generation lost. Needham was very impressed by the soundness—as it seemed to him—of the argument, and went so far as to write a lofty and condescending note telling Spallanzani so.

At this state of the game, the reader might himself attempt to take the part of Spallanzani and—without reading further—attempt to find an answer. It is fairly obvious to a student of microbiology, whose normal working routine is adapted to this form of procedure; but there is no reason why the loophole should not logically be apparent to anyone. Yet it is none too easy to think of at first, and Spallanzani showed his true mettle as an experimenter when he grasped the essential fault of Needham's thesis.

The answer is—as all the best answers are—totally, almost beautifully, simple. He took a series of flasks, corked them, and boiled them for different periods. That was all. On examination a week or so later he found that even those boiled for hours on end were alive with micro-organisms—indeed there were far heavier growths in these infusions than in those that had been boiled for only a few minutes. Spallanzani assumed that this was because the flasks that had been heated for longer had formed a richer broth; infusions boiled for only a few minutes would have released hardly any of the organic matter into the suspending water. He was right, and his undeniable conclusions showed that Needham's experiments were questionable once again.

But then came a further round in the competition. Needham had meanwhile been elected a Fellow of the Royal Society in recognition of his perspicacity and wisdom, a fact which no doubt irked Spallanzani a good deal, and with his new-found academic status he proclaimed that the real failing of Spallanzani's method was that his flasks contained air at a low pressure; and the vegetative principle could not function under those conditions. It needed a raised pressure to function correctly. Spallanzani found that Needham's triumphant prediction was right. Because he had been heating his infusions before sealing the necks of the flask in the flame, the air *was* at a lower pressure than atmospheric. So he carried out his final manoeuvre in the battle. Taking another series of flasks, he drew out their necks in the flame until only the smallest aperture remained. After the period of boiling, he allowed the flasks to cool and only then applied a very small flame to the opening—which was quickly closed. The air pressure inside was therefore exactly the same as the pressure of the atmosphere.

The results of his experiments were unambiguously clear: he had disproved Needham's vegetative force theory once and for all; and he went on to prove that his microbes could not only withstand boiling, but they could survive in a near vacuum too. Even oxygen, he marvelled, was not always necessary for their existence (an observation which Leeuwenhoek had made).

So Lazzaro Spallanzani laid the spectre of spontaneous generation for all time. He settled down to become a noted mineralogist, a great collector for his museum, a personality and eccentric. No-one could doubt now that his views were right; but still nothing happened to the mainstream of biological progress. The minds of men were not yet ready for it. Spallanzani's erudite and outspoken papers (notably his *Observazioni Microscopische Concertenti il Sistema della Generazione dei Signori Needham e Buffon* of 1765) were landmarks of scientific progress, but it was to be a century before Pasteur repeated and elaborated his work and made the results well known.

Elsewhere little was happening. Though the microscope itself was healthily evolving, microscopy was almost at a standstill. A little-known doctor named Boerhaave—a Dutchman—had always insisted on examining his patients' excretions with a lens during his activities in the early part of the eighteenth century and he clearly was the first pathologist on record to do so. But whether it was ever worthwhile is another matter. Later in the same century that prodigious man of medicine and letters, Dr Albrecht von Haller, used the microscope to advantage in his work on anatomy and embryology and by the same time (the middle of the 1700s) Giovanni Morgagni, a noted philosopher-pathologist, had recognized the implications of contagion and refused to dissect bodies infected with syphilis, tuberculosis or smallpox. Yet still medicine did not want to know the truths that lay before its eyes; learned opinion was not yet ready for the nature of infections to become apparent, though by this time there was ample evidence which an intelligent man might have put together into a single concept.

It was not only in the field of microbiology that such conservatism was causing problems. The fate of Dr James Lind and his work on scurvy is a case in point. This disease is due, not to any invasion by pathogens, but to the lack of ascorbic acid (vitamin C) found in the diet of seafarers of that era. The compound occurs in fresh fruit and vegetables, and was naturally lacking in the salt beef and biscuits which featured on the sailing ships of the time. In the 1750s, Lind, a Scottish naval surgeon, wrote several important and quite brilliant works on scurvy. He pointed to the cause of the disease (lack of fresh fruit) and showed that it could be invariably cured by the administration of lemon juice to the sufferer. The evidence he advanced was clear, the results he attained were quite startling. Yet did the medical establishment accept his views?

They may have—but when Lind died over forty years after publishing his famous *A Treatise of the Scurvy* not one of his suggestions had been adopted by the Navy. After his death lemon juice was added to the sailors' diet and then—only then—did scurvy vanish from the Navy altogether. Such are the problems that face the innovator whose audience is unprepared.

Meanwhile John Hunter, the famous anatomist, was adding his own contribution to the study of infection by inoculating himself with pus from a victim of syphilis. Hunter was a brilliant anatomist and became surgeon to the King; yet his education had been imperfect and when he went to Oxford in 1755 at the age of twenty-seven he stayed only two months, saying that they only wanted to stuff him with memory exercises. He spotted these 'schemes' at once, he wrote, and 'cracked them like so many vermin'—sentiments which can be echoed in today's criticisms of much university teaching. He proved that venereal disease could be spread by inoculation and after taking mercuric compounds for three months or so he declared himself cured. In the latter half of the same century, Dr Benjamin Rush in the USA was incredibly close to the discovery of the cause of yellow fever. The condition is spread through the bite of the mosquito *Aëdes ægypti* and though Rush never realized this fact his study of the disease led him to make many suggestions that undoubtedly prevented many worse epidemics in the United States at that time. He often complained about the numbers of mosquitoes, and amongst the measures he insisted upon were the draining of ponds and the clearance of swamp areas—he felt that the disease in some way arose from such regions. He was right, of course; *Aëdes* like all mosquito genera breeds in water. But his observations were based, not on any true understanding of the disease or its aetiology, but on innate common sense and experience of the situation. It was just these very factors which back in England were to lead Edward Jenner to fame as the discoverer of vaccination.

Edward Jenner is a name which we all know. He was, of course, a Fellow of the Royal Society—and quite right too, for someone who had conferred such benefits on society. But in this case as well there is more behind the development than at first meets the eye.

In the first place Jenner was not elected to Fellowship because of his work on vaccination, but for some research in quite a different field. It was in recognition of his discovery that the cuckoo lays the egg in the nest of other birds, and the hatchling ejects the other eggs in the nest as soon as it has emerged. Indeed when Jenner later wrote to the Royal Society on the subject of vaccination he was politely but firmly told to go away and stop meddling in such 'incredible' and 'discreditable' matters.

Smallpox was in the eighteenth century and before as common as measles in Britain and very much more deadly and disfiguring. In 1717 Lady Mary Montagu, returning from the Middle East, advocated that healthy people should be inoculated with pus from a mild form of smallpox and thereby obtain immunity from future attacks. Thenceforth inoculation became standard practice and a fashionistic topic. Meanwhile in the country it was widely known that cowpox—a similar though markedly milder disease—seemed to induce immunity too, and it was common for country people to try to catch cowpox themselves in order to avoid the ravages of smallpox. Clearly the answer was not hard to find—it was to inoculate people, not with mild smallpox at all, but with cowpox. And it fell to the lot of Jenner to try it out experimentally for the first time.

In a foolhardy and dangerous experiment he inoculated a

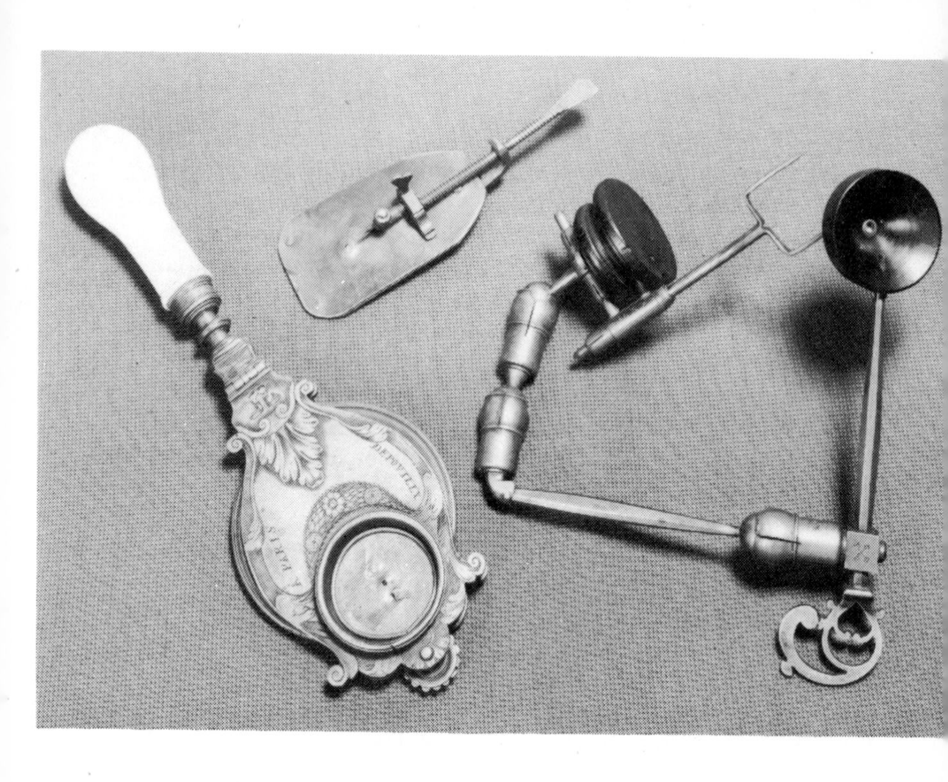

Three early designs of simple microscope.
Top. **Leeuwenhoek microscope of the typical form.**
Lower left. **Engraved French microscope. Note circlip holding metal disc with small lens at its centre.**
Right. **Samuel Musschenbroek was a pioneer microscope-maker of Leyden and his younger brother, Johann, made this example—similar microscopes were made especially for Swammerdam. Science Museum Photograph.**

young boy of eight with cowpox and then, later, scratched some smallpox pus into his arm. The boy remained free of the disease. In 1798 he published a book on the subject which provoked widespread controversy and debate, and by the time he died in 1823 vaccination was widespread. So we see the new heterodoxy becoming orthodoxy in a few short years; and since then—as we may also see elsewhere in science and society—orthodoxy soon becomes conservatism. Thus, when Jenner first announced his 'discovery' (*affirmation* would be a better term, surely) he was greeted with undue scepticism by scientific society:

Development of the microscope.
Left. **Typical microscope made by John Marshall around 1700. Compare with Hooke's instrument (page 35).**
Centre. **Microscope by John Cuff, which in 1744 effectively founded the lacquered brass tradition.**
Right. **George Adams's 'Universal' microscope was probably made a decade later, and has something of the appearance of a nineteenth-century instrument. Science Museum Photograph.**

unable to doubt their preconceptions, orthodox scientists were instinctively opposed to his views although they were based on a widespread popular belief, with plenty of incidental and experimental material behind them.

But once vaccination *had* become accepted, exactly the opposite state of affairs prevailed. Now Jenner had become a legendary saviour of mankind, a clinical father-figure with the kind of status quite out of proportion to his attainments. And this has persisted to this day. One recent writer quoted Jenner's words on the subject:

> May I not with perfect confidence congratulate my country and society at large on . . . beholding . . . an antidote that is capable of extirpating from the earth a disease which is every hour devouring its victims; a disease that has ever been considered as the severest scourge of the human race. . . .

Surely the most self-congratulatory sentiments ever expressed by a discoverer about his findings. Yet, so imbued are we with the almost superhuman and saintly qualities of this inquisitive countryman that the writer concerned described these very words as being 'mild, modest phrases' in all seriousness. Nothing could feasibly be further from the truth.

And so we may observe another phenomenon in the affairs of mankind which has exerted a profound effect on scientific endeavour and the pattern of progress. Not only is conservatism a stolid and unyielding block to development, for no discovery, no matter how well founded, can ever succeed in converting the establishment if it is not ready to be converted, but the eventual effect of continual repetition of an idea may suddenly result in its becoming popular or, as we may say, fashionistic—and once that has happened everyone will tend to leap on the bandwagon; the individual concerned may change from being a man for derision (which he did not deserve) into a genius of rare talent (which he may not warrant either) simply because of the whims of establishment opinion. This can prove to be the greatest of all obstacles to progress.

So we move towards the end of the century. On the technical front the microscope had reached considerable levels of sophistication: the rotating multiple nose-piece seen in modern microscopes had been devised, the mechanical parts of the instrument had become refined and perfected to a rare state of *finesse*, even optical requirements had been vastly improved. Yet the microscope as a research tool was still largely ignored. Spallanzani had certainly used it to advantage (though the microscopical observations he made are useless from the point of view of identification), and so, as we have seen, had others; but their work had been overlooked in the main, and where it was not overlooked it was soon forgotten. Most surprising of all is the welter of cir-

Adams's design for a 'Lucernal' microscope was published in 1787, in *Essays of the Microscope.* The Argand lamp *(right)* was the illumination source, and the device—labelled LM—on the far left was an eye-guide. Looking through the aperture in this, when the microscope had been correctly adjusted, gave a very bright field of view. Alternatively a ground glass screen placed at this end, in place of the eye-guide, enabled a number of viewers to watch—just like colour television.

cumstantial evidence at that time which suggested a microbial theory of disease—and it was this revelation which was to be the microscope's greatest benefit to mankind. Lind's work, though not connected with a micro-organism of any kind, did at least introduce the concept that diseases could be eradicated by suitable means: that was, in its time, a sensational conceptual development.

Rush came desperately close to discovering the vector of yellow fever and, as we saw, uncovered all but the last link in the sequence of events leading up to the spread of the infection; Morgagni had assumed that diseases were infectious, and Hunter had shown it to be so by self-experimentation. Spallanzani's work had dispelled the myth of spontaneous generation and had shown the hardy nature of many micro-organisms, and of course there were Leeuwenhoek's classical revelations on the ubiquitous nature of microbes of all kinds to fall back on. Underlying it all was the time-honoured idea that diseases might be due to micro-organisms, a notion which as we have seen was current before the microscope ever materialized as such and which was later revived, independently and cogently, but which was denied widespread acceptance. Anyone who had taken hold of the problem in all its implications, who had surveyed the literature and had then sat back to think it over in full, could hardly have failed to deduce that there were important implications for mankind in general and for science in particular.

The lacquered brass and silvered microscopes of the period became party pieces, gaudy toys to reveal something spectacular yet uncomprehended in a droplet of stagnant water; trinkets of a materialistic society. But, though the innovators had been relatively few and their impact had been slight, the repercussions were still there. The ripples were still spreading—even though it was going to be half a century or more before the effects began to manifest themselves.

4 The Truth Begins to Dawn

USERS of microscopes for entertainment and study in the eighteenth century continued to find one important factor of considerable annoyance: the lenses tended to produce rainbow-hued fringes around the object under study. The images were coloured, or chromatic, and this effect detracted greatly from the clarity of the result. The reason for this is elementary physics. To elucidate the mechanism further we must first examine the way in which light behaves as it passes from one medium to another.

It is an old saw that 'light always travels in straight lines'. But it does not, of course. Not only are there minor discrepancies due to gravitational pull (it was investigations of Einstein's laws which demonstrated that) but rays of light are continually bent or refracted slightly by even the changes of temperature from one part of the atmosphere to another. So familiar are we with this phenomenon that it is surprising how unaware of it most people seem to be. The haze of heat rising from a road in high summer, the hot air curling from a radiator, even the column above a lit candle—all of these produce density changes in the air and corresponding movements in light rays passing through. Stars twinkle for the same reason, as their light passes through the uneven, turbulent atmosphere above our heads. The nature of the effect is plain to see in the diagram (page 92); when a wave-front of light strikes a medium which is optically more dense (when it moves from air into glass, for point of argument) the whole beam slows down.

But if it strikes at an angle the effect of this slowing is to turn the beam to another direction—it is bent so that it enters the glass block more nearly at right angles (i.e., it is refracted *towards the normal*). This is the way in which a lens works. And at once there are two sources of optical error or aberration which may be introduced.

The first is spherical aberration, which as we saw earlier was

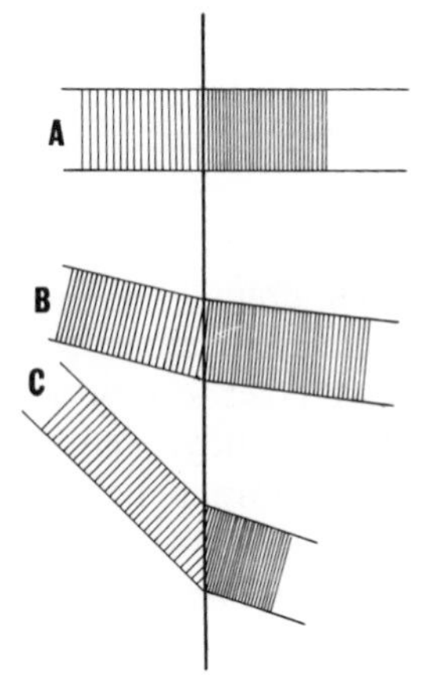

Light entering glass as in (A) undergoes a change in speed; in (B) and (C) we see how this effect causes refraction.

a fault of the early bead lenses. The surface was roughly circular, and light from the periphery was brought to a focus nearer to the lens than light which passed through the central regions.

The result was predictable. If an image was placed so that its centre was in focus, then the edges would be totally blurred; conversely if the outer details were focussed the middle was indistinct. It is possible to reduce the error by stopping-down the lens (i.e., by restricting its diameter) but this reduces the amount

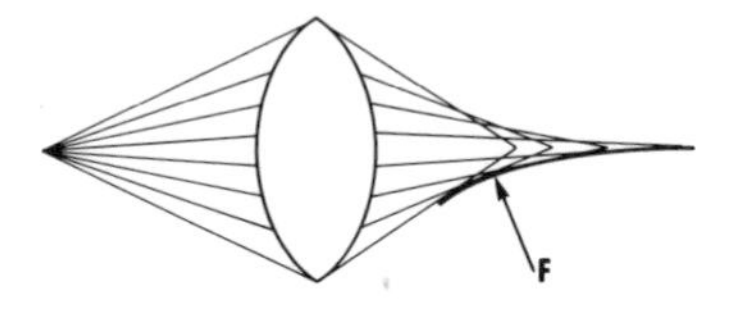

A spherical lens causes spherical aberration.

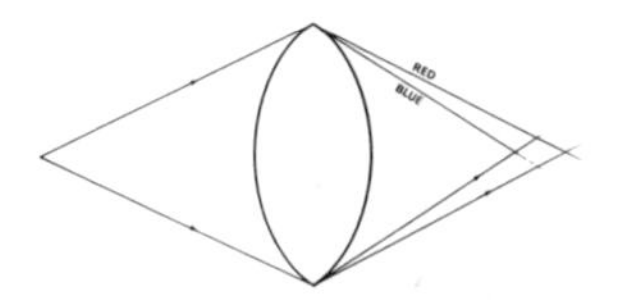

Chromatic aberration: red light comes to a focus beyond blue.

of light to such an extent as to make the magnified image very dim. Otherwise the profile of the lens may be altered and specially contoured so that the rays come to focus in the same plane. This is what Leeuwenhoek managed to do by his delicate grinding technique.

But this is still not enough. For it is equally fundamental for light rays of a greater wavelength to be refracted less than those which are shorter—i.e., red rays are refracted considerably less than blue. This means that a single lens which is corrected for one colour will inevitably be uncorrected for all the others—and hence the second deficiency: chromatic aberration as it is termed.

In the seventeenth century it was widely felt that the error was uncorrectable by definition. Newton had written that the amount of dispersion (i.e., the spread of a spectrum) was a function of the amount of refraction to which the ray had been subjected. Newton was seen as an all-wise, all-knowing figure and the conservatism that resulted made it impossible for anyone to seriously question anything that Newton had written. That was unfortunate, as this belief was based on empirical knowledge and Newton naïvely accepted it without carrying out the tests that he should have. In any event he was wrong. Chromatic aberration could be overcome. One method he foresaw was to do away with lenses altogether and use mirrors instead. Several microscopes were designed (amongst the first being those of Barker and Smith, independent British workers who were first active in the late 1730s) which were rather like small reflecting telescopes with lenses as eyepieces. But they were preliminary. The first reflecting telescope had been designed by Gregory back in 1662, and ten years after that Newton had published a note in *Philo-*

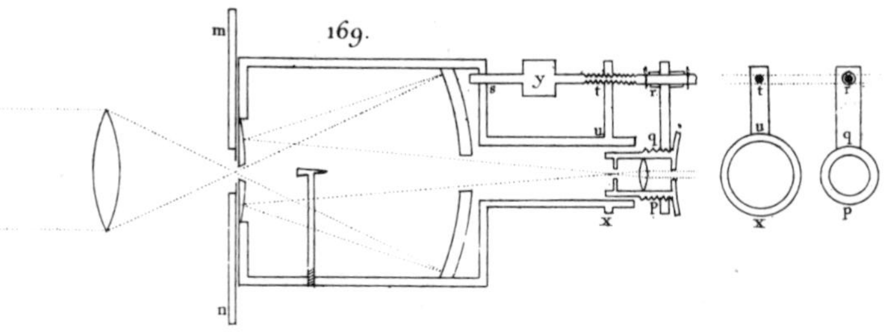

Smith's proposal for a reflecting microscope design, intended to overcome chromatism in the image. From *A Compleat System of Opticks* (1738).

sophical Transactions which suggested that reflecting microscopes would be the only answer to chromatism.

It appears that the first man to find a practical answer to the problem was a lawyer named Chester Moor Hall, who realized that glass lenses of different densities might be used in pairs to correct each other. He made one convex lens from crown glass and another matching concave lens (lower in power than the other) of the denser flint glass—the variety used for high-quality cut-glass today, so high is its refractive index. In making the lenses he was obliged to seek the help of professional manufacturers and so he sent the lenses separately off to two London specialists under conditions of secrecy. However it seems that they both sub-contracted the work to the same grinder, and so the nature of the experiment was revealed. Moor Hall was certainly the earliest recorded individual to produce an achromatic lens, though he did not apply the principle to the microscope. Inevitably, within a score years or so, telescope lenses made to this design were becoming common. It seems that in 1774 an achromatic lens was supplied as a standard fitting to a British solar microscope, though it was intended for projection on a screen and not for observation in the ordinary way. And there are records which suggest that a Dutch father-and-son team (Jan and Harmanus van Deiyl) constructed a microscope achromatic lens system around 1770.

Certainly by the early part of the nineteenth century there were many manufacturers of low-power achromatic lenses, and by the end of the 1820s they were successfully used by many workers as a standard item of equipment. This improvement in lens quality was timely; it enabled microscopists of the period to dispel one important concept which, though fanciful and erron-

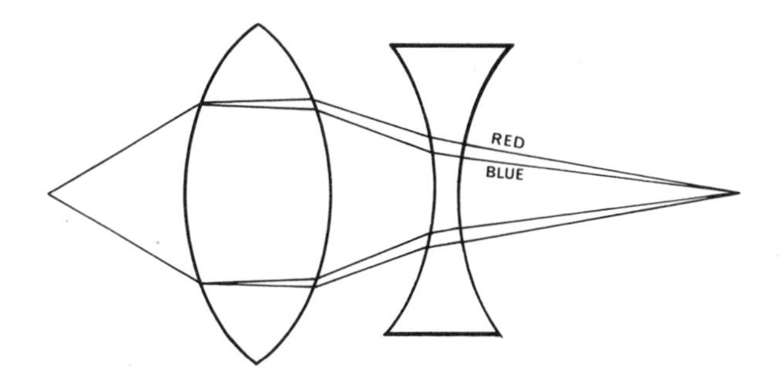

Positive and negative lenses correct the disparity in focus between red and blue light.

eous, was almost a fore-runner of the cell theory. It was the idea of 'globulism'.

This had arisen out of the failure of early lenses to resolve fine detail. Small, bright points were seen only as hazy circles, and the use of sunlight especially encouraged workers to stop down their lenses and see any fine detail surrounded by a diffraction fringe—a circular halo of light. These phenomena were identified as being 'globules'—particulate entities common to many (if not all) forms of matter, and believed by some, no doubt, to correspond to atoms—and the 'globulist' theory became widespread. It was the use of achromatic lenses by workers such as Lister (the elder) and Hodgkin in the late 1820s which showed that 'globules' observed in many tissues were non-existent. They doubtless had a part to play in the so-called homunculus theory, the belief that there existed in every sperm a tiny adult human being. Indeed, if I point out that many 'globules' proved eventually to be cells, and the coded genetic material in each sperm head *is* a blueprint for all the details of a mature adult, a point elaborated on page 150, we can see that these theories were perhaps less absurd than we care to admit.

Some years earlier, at the turn of the century, a young French surgeon and physiologist had come close to founding the science of histology—the microscopic study of tissues. He was Xavier Bichat, whose most notable work *Recherches physiologiques sur la vie et la mort* was published in 1818. This work began with the words: 'La vie est l'ensemble des fonctions qui résistent à la mort.' This has generally been denounced as a truism, a naïve statement of the obvious, an example of begging the question. But it is nothing of the sort. Bichat was implying that death is the 'normal' state for any chemical system (which the body, of course, is) and that life was a continual effort to resist death. He is right, and his view was a philosophical departure of considerable validity. It suggests that only the continued actions of man and his constituent parts prevent morbidity from taking over. Perhaps the most telling demonstration of this thesis is in the modern use of immuno-suppressive drugs in transplant surgery. The effect of these compounds is to prevent the tissues from fighting against foreign protein which may have gained access to the patient's body; it is not in any way a deleterious propensity *per se* but simply the removal of one of the body's means of 'résistance'. Disease is likely to follow, and death supervenes as a rule. Again, the use of anticoagulants invariably results in difficulties for the patient, and at least one rodenticide comes into this category. It kills, not by specifically introducing any toxic element into the system, but by the removal of one of the body's 'defensive functions', as Bichat would have called it. And, since we can only resist disease and death by continued physiological defence reactions, the results are potentially serious.

Happily, Bichat's philosophical nature extended into other fields of work. He was obsessional about research, often spending the night in the morgue busy in dissection, and during one winter term he is reputed to have performed over six hundred necropsies. He used the microscope as an aid to the study of tissues in health and disease, and his attention to pathological processes on the minute scale set a precedent which medical research was destined to follow and which is a mainstay of medical investigation today. Yet he knew the limitations of his instruments and did not trust lenses. He died at the age of thirty-one in 1802, just as the microscope was about to reach a degree of technical perfection compatible with his projected investigations. Who knows what he might have done?

It is strange to reflect that a compatriot, Jean Cruveilhier, worked in much the same field some thirty-five years later and produced a remarkable study entitled *Anatomie pathologique du corps humain*, the illustrations in which are amongst the best ever produced. Cruveilhier, however, did not use the microscope at all.

At about the same time, in 1837, an Irish doctor named Abraham Colles (of 'Colles fracture' fame) published a paper on venereal disease and observed that a child born of a mother without any obvious signs of syphilis but which shows the disease itself when a few weeks old may well infect others, but will not infect the (immune) mother—an observation known to this day as Colles's Law.

And another great pioneer in the study of diseased tissues was Thomas Hodgkin, demonstrator at Guy's Hospital, who first described Hodgkin's Disease—lymphadenoma—but who did not use the microscope either. Which is a pity, for several of the cases he described were not of 'Hodgkin's disease' at all.

But suddenly, all this changed. Overnight almost Dr Theodor Schwann was to begin a systematic study of the microscopic structures of plants and animals and in 1839 he published his legendary *Microscopische Untersuchungen über die Übereinstimmung in der Struktur und dem Wachstum der Tiere und Pflanzen* which was the first work of histology ever published. And two years earlier, as if in testimony to his grasp of the whole subject of microscopy, he had published a notable paper which was an extension of Spallanzani's notable investigations. He discovered yeast as an agent of fermentation, and carried out a number of experiments which led to conclusions such as: a boiled organic substance will not ferment if air introduced to it has been heated; unboiled or unheated air or organic matter must be present for fermentation to commence; a plant—a 'sugar fungus'*—causes fermentation in grape wines; chemicals which are toxic to infusoria but not to fungi do not inhibit fermentation; and so on. Many of these conclusions were reached independently by the French scientist Charles Cagniard-Latour, who studied yeast microscopically and showed it was able to withstand freezing and to grow anaerobically (i.e., without oxygen).

But meanwhile the renowned chemist Justus Liebig published, in 1839, an incoherent and pedantic account of fermentation as though it were a purely chemical phenomenon. There was not a breath of experimental evidence in the account, which by standards contemporary with his was rambling and dogmatic. But

*The generic name of yeast—*Saccharomyces*—is a literal translation of this term.

Liebig had a substantial reputation, and so acceptance of the clearly-demonstrated experiments of Schwann and Cagniard-Latour was postponed for a generation or so.

These papers on microbiology were amongst the first to result from the widespread acceptance of achromatic lenses, and it was only a matter of time before the sudden realization of the significance of it all became apparent. The establishment was patently receptive to such ideas, after the continuous wearing away of prejudice which these papers had brought about. The ripples had spread, the academics were attuned to such revelations and the experimental evidence (together with that of their own senses and experience) was undeniably convincing. Not only that but the optical quality of microscopes was vastly improved. The whole subject of bacteria, diseases and the cell theory was poised like an egg balanced on the bonnet of a bus. Surely now the mystery would shatter, the truth spill out?

Yet still it did not happen. In 1844 a largely unknown biologist named Agonisto Bassi published a treatise entitled *Sui contagi in generale* in which he expanded some earlier research he had carried out into the nature of a fungus disease in earthworms. A truth had dawned on this forgotten Italian scientist and he wrote that many diseases 'including smallpox, spotted fevers, bubonic plague, syphilis' were caused by 'living parasites whether animal or vegetable'. No notice was taken of this, either.

Perhaps the last man to suffer from the non-acceptance of microbiological principles was Ignaz Semmelweis, physician to Vienna's Obstetric Hospital, who in 1847 found the great benefits of the use of chlorinated water as a hand-rinse solution for doctors in obstetric wards. In repeated trials he showed how the incidence of puerperal fever could be drastically reduced by the use of this antiseptic solution, but he was repeatedly shunned and the idea was continually ridiculed. There was not a scrap of reason behind the non-acceptance, and all the data available showed that he was undeniably right. But when he died, in 1865 at the age of forty-seven, he was a frustrated and depressed man. Not a word of his doctrine had been accepted. How different it was when Lister announced similar findings twenty years later.

The snowball began with Rudolf Virchow, a small unprepossessing man of medicine who in his thirty-seventh year announced a theory which has underlain microscopical research ever since. His theory was not an original construction, coined from the thin air of intellectualism; rather it was the pooling of the many ideas that had been published before. It contained Hooke's original references to cells, Leeuwenhoek's studies on free-living protozoa, Spallanzani's observations of binary fission, Schwann's revelation that tissues were composed of cellular entities. To this pattern of conceptual knowledge, Virchow added a new factor which drew the net closed: cells were formed by the reproduction of other cells.

At once this key to the nature of life, coming as it did on the heels of persistent propaganda dating back for several centuries, soon swept across Europe and beyond. At one fell swoop it was clear how animal and plant species grew and matured, how ova functioned, how embryology worked; it became obvious how the disease process might function, how man himself was made; the understanding of the role of the cell suddenly brought about a synthesis of ideas from many disciplines. At last the different paths had met.

Virchow published his theory in 1855. Just two years later another scientist entered the scene, a chemist by training—Louis Pasteur. In his paper *Mémoire sur la fermentation appelée lactique*, dated 1857, he wrote of his microscopical observations on fermenting liquors. The quality of his work was no better than that of Schwann and Cagniard-Latour, on balance, but of course it came at a fashionistic period in history. In decrying Liebig's theory that fermentation was a chemical process (the earlier work of these other two microbiologists having by now been largely forgotten) Pasteur was riding on a wave of acceptability. The microscope had gained currency as a scientific tool, rather than an instrument for the drawing-rooms of the nobility; and Virchow's convincing demonstrations were entirely compatible with what Pasteur was saying. Not only this, but in his twenties Pasteur had made the first of his notable discoveries—which do not concern us here—in the chemical field, by observing stereo-isomerism in tartaric acid (the property of apparently identical molecules to form as mirror-images of each other). Pasteur was extroverted, a friend of the press, a positive headline-seeker. De Kruif wrote of Cagniard-Latour, 'He was no propagandist, he had no press agent to offset his modesty.'

Pasteur was quite the opposite. He was always keen (as Jenner, for instance, had been) to blaze his trail in hyperbole. When he observed micro-organisms surviving without oxygen he was quick to label this a 'formidable' discovery, forgetting about Spallanzani's discovery of the same phenomenon a century earlier, and Leeuwenhoek's a century before that.

Pasteur then set off to carry out a long series of experiments in which he heat-treated broths and infusions, and kept them sealed out of contact with the air, showing how they remained uncontaminated by any microbial growth. For these experiments he became in his day—and is now—famous, yet they did naught to improve on Spallanzani's work. Except for one thing: Pasteur carried out an ingenious investigation in which, so as to allow untreated, unrestricted air to gain access to his heat-treated liquids, the atmosphere was admitted through a long, fine, curved tube. Any particles were unable to move along it since there were no currents of air to carry them. This was, it is true, the first time that broths and infusions had been kept sterile whilst in contact with the air; and it was clearly a brilliant new move in the argument.

But, though the fact has been widely overlooked, this 'breakthrough' idea was not Pasteur's at all, but was the brainchild of Balard, discoverer of bromine. Later Pasteur tried a further experiment, childishly simple in its execution, in which he opened flasks that had been heat-treated; some were opened in a cool cellar, others in the dusty open air. And he found, as one would expect, that the micro-organisms were more abundant in dusty air—that they were carried in dust, in other words. Though this was only a further refinement of Spallanzani's methods, it was something of a personal step, for this was the first of Pasteur's microbiological experiments that was really his own.

Later, he became embroiled in a public controversy (Pasteur was a master of the public demonstration, surely the forerunner of today's press conference) because, whereas he had opened flasks of infusion high on Mont Blanc and found the air sterile, three adherents to the spontaneous generation theory, Joly, Musset and Pouchet, had found their own infusions to be swarming with bacilli. Pasteur called them fools, they called him a bigot; but in fact (within the bounds of contemporary knowledge) they were both right. Some spores of bacilli are well able

The amateur microscopist of the second half of the nineteenth century often used a specially-constructed wooden stand for the instrument. Most of the great microscopes in vogue at the time were too tall to use in the vertical position if they were placed on a table-top, and this upright mode is obviously important if liquid preparations—such as pond-water—are to be observed successfully.

to withstand prolonged boiling; it was these which had contaminated Joly's, Musset's and Pouchet's flasks. Though Pasteur was able to prove that he was right, he was quite wrong to assume that this necessarily meant the other three were wrong. This failing, the assumption that because a statement is true another must be false, still dogs the heels of modern scientific argument. It was years before the answer to this particular riddle emerged, and for the time being it merely confirmed Pasteur's reputation as an invincible and original pioneer. Neither of these descriptions was accurate.

Though it would be wrong to credit Pasteur with too much originality it would be equally erroneous to overlook the fact that he was, without doubt, a clever and ingenious man. In the mid 1860s he worked in the South of France on diseases of silkworms, and established that the condition was contagious; he outlined methods of better hygienic practice which helped to control the disease and observed the causative organism too. This was to start something of a battle between Pasteur and the establishment once again, and showed a serious weakness in Virchow's theories.

In elaborating his thoroughly sound cell theory, Virchow had carried it a little too far by suggesting that disease was a chemical deviation from the normal state—a theory essentially comparable to Liebig's views on fermentation. Large numbers of scientists, including Ludwig, Helmholtz and Brücke, agreed with this point of view, so popular had Virchow's other work become. But Pasteur felt in his bones that fermentation and disease were similar processes—not chemical, however, but microbial. He was determined to prove the point.

There were precedents here too, however; in 1840 Henle had published a theory that disease was due to the intake of some morbidity principle, and he explained the incubation period of a disease in these terms. From translations of his work it is clear that he took the analogy of microbiological fermentation very seriously; and this anticipated much later work in the field of bacteriology. Though Semmelweis introduced antiseptic hand-rinse solutions a little later (page 98) he did not, of course, owe this to any clear understanding of the nature of disease but rather to the basis of experience; and Henle was clearly breaking new ground with his theories. Ten years later, in 1850, the French scientists Rayer and Davaine smeared some blood from cattle infected with anthrax on a microscope slide and observed small rod-like organisms. They were observing the bacillus which causes the disease. Later developments were muddled and contradictory, for they failed to establish any direct cause-and-effect relationship between the bacteria and anthrax itself. Some workers examined blood from cattle and missed the bacterium altogether, whilst others infected animals with contaminant organisms so that they died of some other disease. But Davaine pressed on, arguing, lecturing, demonstrating and corresponding. He examined blood from an infected cow and that taken from her foetus, showing that the placenta kept the bacteria at bay.

He filtered serum samples through unglazed china and showed that the filtrate could not cause the disease because it contained no bacilli; in short he all but proved the cause-and-effect relationship. The problem of the persistence of anthrax was still to be solved, however, and it fell to the lot of Koch—an unsophisticated country practitioner—to locate the missing link. This was the formation of spores by each bacterial cell: hard, encapsuled bodies which could withstand desiccation and so lie dormant in soil for many years. As we shall see, this splendidly inventive and hard-working man (who carried out much of his work independently, in his own laboratory) made many important discoveries and demonstrated the occurrence of bacteria specific to many diseases. He also found methods of cultivating bacterial growths on solid culture media—the technique most widely used today. The credit for a good deal of Pasteur's work might better be given to Koch: yet even he had important influential (yet neglected) precedents.

One of the principal researchers at that time was undoubtedly Edwin Klebs, born in Königsberg in 1834. He worked after graduation as assistant to Virchow, and in 1871 he made some microscopic studies of gunshot wounds where he found a variety of micro-organisms. He did not recognize that they belonged to different genera, and lumped them all together under the heading *Microsporon septicum* as though they were all one species. Koch's similar work came later than this, about 1876; perhaps he was stimulated by Klebs's research. It was Klebs who first made a solid, jelly-like culture medium for bacteria by adding fish-glue to a broth and allowing it to set into a semi-solid jelly. This work dates from 1872, and it was nine years later that Koch perfected a similar process which utilized gelatine as the gelling agent instead. Growths on solid media are the basis of the pure culture, since without some firm base on which organisms can grow, different species may become inextricably mixed—it is pretty certain that Pasteur used impure cultures for much of his work, for instance—and this is clearly an obstacle to detailed investigations. Growths on jelly-like media occur as raised circular colonies, each descended from a single cell, which remain (usually) discrete and distinct. Since many cultures are best incubated at blood heat (37°C) and gelatine media melt at this temperature, the modern bacteriological culture media contain agar, a seaweed extract which only liquifies near boiling-point.

Klebs established long-forgotten precedents in other fields too; it is recorded that he produced syphilis experimentally in a monkey by inoculation in 1878, an experiment generally first credited to Metchnikoff in 1905. And he saw the bacteria of both diphtheria and typhoid fever before the 'popular' discoverers, Löffler and Eberth respectively.

But Klebs lacked the out-going personality of men like Pasteur and his work was never properly recognized. He was, to be realistic, a bit of a dabbler in some respects, and did not carry a discovery through to its logical conclusion once he had reached a point which satisfied him. None the less, he is deserving of far greater credit than it has been customary to grant him.

Koch, still working quietly away as a devoted, enthusiastic independent scientist, had still to make his greatest single contribution. It was very interesting to demonstrate these different species of bacteria in wounds, in septic blood-samples, in swab samples taken from diseased men and women; but how may we be sure that the organism and the disease have a cause-and-effect relationship? In recent years we have seen bacterial bodies demonstrated in the joint fluids of arthritic patients; 'perhaps this is the causative organism,' infer the doctors. We have seen the same occur in cases of cancer in the past, too, and a much-publicized episode some years ago centred on a charlatan who had demonstrated small particles in leukaemic blood. But they were not the cause of leukaemia either. More recently we have seen cigarettes and lung cancer spoken of at government level as having a direct cause-and-effect correlation; but has anyone demonstrated it experimentally? No. It seems certain that cigarettes have something to do with the likelihood of contracting the condition, but more cannot scientifically be claimed. Recently we have seen moustaches and sexual repression associated; a tendency to lisp linked with concealed traits of a homicidal nature; myopia with intelligence—the list is endless and at times amusing.

So how can the field of bacteriology, new, exciting, all-inviting, have seemed to Koch in the 1870s and beyond? Everywhere were bacteria, some small and round (*Cocci*), some elongated and sausage-shaped (*Bacilli*), others spiral like a corkscrew (*Spirochaetes*).

He found them everywhere, many in perfectly healthful environments. It would have been easy to observe bacteria consistently in association with diseases (as—in other fields—we are doing today) and to assume at once that this was the causative organism. But Koch was a brilliant man all the more for his faculty of self-criticism, and he devised methods which allowed him to check and recheck the thesis before he accepted it. Eventually in a paper devoted to the identification of the organism which causes tuberculosis (*Mycobacterium tuberculosis*) he outlined his basic teaching. The four steps to certainty are now known as Koch's Postulates, and they are:

(1) that the organism must be observed in every case of the disease;
(2) that it must be isolated and maintained in pure culture;
(3) that inoculation into a susceptible animal must produce an identical condition;
(4) and that the same organism should be isolated from the experimental subject.

His work on tuberculosis was the start of a flood of research, now that the techniques were at hand;* within the next few years (the paper was published in 1883) a vast range of bacteria was isolated and most of the bacterial diseases were in the space of a decade or so investigated and their specific bacteria isolated. Some diseases (such as pneumonia and meningitis) have since proved to be more complex than was at first suspected, as several causative organisms—viruses amongst them—have been found. But the flood-gates had been opened, and by the end of the century the role of bacteria and their allies in the causation of disease was very well understood. It was, without doubt, an exciting and unique phase in microscopical research. And it was perhaps the most important single period in understanding more

*The closely related organism of leprosy, *Myco. leprae*, had been discovered in the early 1870s by the Norwegian Hansen, but the work had attracted little or no attention.

about the invisible universe since Leeuwenhoek's active period.

Why is it, then, that it is the name of Pasteur which stands out above all others when we think back to this period? There are several reasons. Some of them throw a light on the latter-day fame of some of our contemporary men of science and medicine.

In the first place, he did carry out some very considerable original research. Much of his work in the bacteriological field was borrowed from precursors or contemporaries but some of it was undeniably original. His investigations into the diseases of yeast which produced undesirable fermentations in the French wine and beer industry, correctly attributed the cause to several species of bacteria, although his culture methods were imprecise and greatly inferior to those of Klebs and Koch. In particular, by carrying out experiments based on an analogy with vaccination, he was able to produce vaccines against anthrax and, later, to produce anti-toxin preparations against other diseases. This was sound, original research—though it was, once again, based on that of others; it was certainly not as novel and systematic as work by some other scientists. This research programme cannot in itself explain Pasteur's fame.

But, secondly, Pasteur carried out work in a field close to the heart of the common people. He was not to spend long periods in continuing the abstruse chemical research of his student days; he preferred to deal with beer, wine, silk factories, and to examine the main 'scourges of mankind'. Thus his investigations were orientated towards the public: and his discovery that the heating of wine to a temperature of around 60°C (well below boiling point) would destroy contaminant bacteria was quickly named pasteurization. His name became, through this, almost a household expression. A century ago he was working in the Whitbread brewery in London—and what is nearer to the working man's heart than his pint?

Thirdly, he clearly gave considerable attention to his public image. He was a master of the considered public statement, and he perfected the press conference in his time. For example, when he had devised a means of weakening the rabies virus in experimental rabbit material so that it could be used as a therapeutic vaccine, he was clearly eager to try the experiment in a human patient. On 6th July 1885 a young boy of nine was brought to him with the bites of a rabid dog on his legs and body. Here was

Pasteur's opportunity—without help the boy was virtually certain to become very ill and to die.

Pasteur's laboratory work had shown beyond reasonable doubt that his vaccination material, prepared from the spinal cords of rabbits, would work; and knowledge showed how certain the boy was to die. Nothing was to be won by refusing to treat him and the evidence suggested that everything was to be gained by doing so.

With the help of two colleagues, Doctors Grancher and Vulpian, he carried out the vaccination. Subsequent accounts are heavily laced with comments on Pasteur's need to be 'urged' before he would carry out the experiment, and his apparent 'agitation' and concern. Excitement and enthusiasm he must, surely, have shown; but the other comments are almost out of character and have something of the ring of the student who, whilst waiting for the results of an examination which he is confident he has passed, confesses to an uncertainty which he does not feel. It is almost a stock in trade of the scholar. Would it not be fair to suggest that this extroverted and confident man was preparing the public for the success he was sure would come, whilst insuring against the unlikely eventuality of failure at the same time? The circus juggler often drops a club, quite deliberately, before the highlight of his act in order to remind this audience how truly difficult it is. Pasteur was a conceptual juggler, and a performer in public; so why should not his approach have been similar? In any event the news of the successful outcome of this first trial in a human subject spread like wildfire. Pasteur had done it again.

Not only were some of his experiments handled in this way, but Pasteur became a master of the demonstration before the press. In 1881 he had perfected a vaccine against anthrax. The disease was costing France tens of millions of francs per year at that time, and many scientists would have been pleased and eminently satisfied to elucidate a means of prevention. But as we have seen, it is not the merit of a discovery which matters foremost in its acceptance, but its presentation. Had Pasteur been content merely to publish the development it might have attracted too little attention, and he was keen not only to make the discovery but to publicize the fact as well.

Accordingly, as soon as the laboratory tests had shown the

efficacy of the technique, he summoned the press and public to a field at Pouilly-de-Fort on 31st May 1881 and proceeded to explain and demonstrate his method, announcing that the twenty-five vaccinated animals would be protected from anthrax, and the same number of controls would succumb. He then carried out the injections, and asked everyone to return on 2nd June to witness the truth and accuracy of the prediction. Return they did, if anything in greater numbers, and Pasteur's prediction was so utterly clear to all that there was no limit to the adulation he was shown.

And he had, at the same time, demonstrated the necessity of conscientious public relations in advancing any new discovery. We do not have to look far to find more recent examples of the same approach.

So, by the end of the century, microbiology had truly come of age; the unseen, unsuspected micro-organisms that swarm through us and our environment—on whom our very way of life depends—had suddenly been recognized as potential enemies of mankind. This point of view persists today, unfortunately perhaps since the overwhelming majority of microbes of all kinds are either innocuous or vitally necessary to life as we know it. Their role in decay, in the production of atmospheric carbon dioxide, in the processes of digestion—they are varied and fundamentally important and it is basically wrong to imagine that 'bacteria' and 'germs' are synonymous. Yet the public understanding tends to do exactly this—and as we have seen a preconception, once gained, is extremely difficult to overcome. Our attitudes today stem directly from this period of bacteriological investigation before the end of the last century.

What had been happening in other allied fields? The study of chemical photography had been advancing steadily; in 1836 it is recorded that J. B. Reade took the first primitive photomicrographs* with a solar microscope; and two years later Fox Talbot took some pictures of mounted specimens which are remarkable for their clarity.

*A photomicrograph is a picture of a specimen taken through a microscope; a microphotograph, conversely, is a very small photograph (such as was used in the pigeon post, or more recently in microdots); but the latter term is frequently misapplied to the former definition.

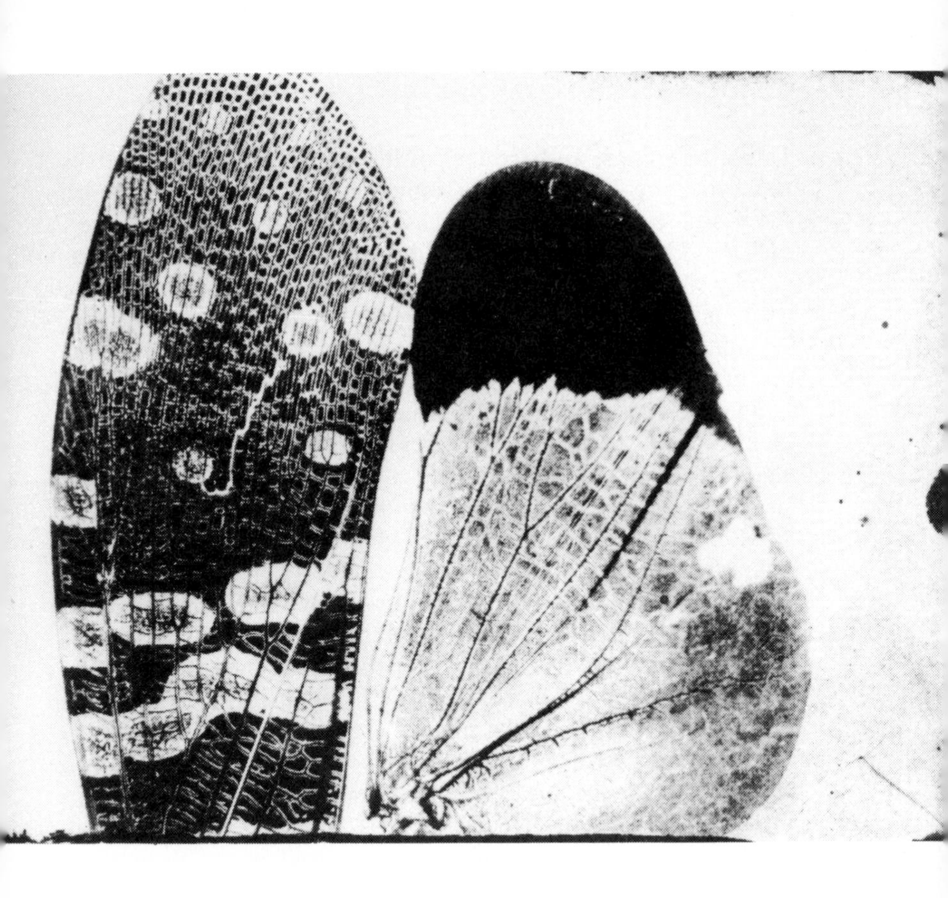

An example of the first photomicrographs ever taken. Dated 1841, this view of coleopteran wing structures (identified at the time as 'butterfly wings') is a study by Fox Talbot. The original is a browned negative on absorbent paper. Science Museum Photograph.

These lay in folded black paper amongst the collection of Fox Talbot equipment in the Science Museum, London, forgotten for some years. I recall the first time I saw them; they were no more than torn squares of blotting paper bearing the faded brown images of his first experiments in this field. I carried out almost identical experiments myself at the age of sixteen, not knowing anything very much about photographic chemicals but being aware (as he had been) that silver nitrate tends to blacken paper when exposed to sunlight. I can identify with Fox

Talbot's feelings, and the feel of his primitive paper negatives was evocative—but then his results had been very much more successful than mine as well as being more than 120 years earlier! Photomicrography had developed alongside photography of the more conventional variety until it had become used as a tool of research. Koch's studies of bacteria are legendarily detailed and constitute fine examples of nineteenth-century micrography.

The microscope itself had continued to develop too, of course. One of the most important changes had been the incorporation of a condenser system beneath the stage on which slides were laid. It had become increasingly well documented that the use of lenses to concentrate the light in this way would improve the quality of the visual image, but—particularly on the continent—it was generally believed that, no matter what the theoreticians might say, condensers were superfluous and even damaging to the results. Eventually this tradition died and condensers became standard fittings to microscopes with a considerable improvement in the quality of the image. Traditional designs, such as the drum microscope popular at the beginning of the nineteenth century, gradually modified themselves until they closely approached the present-day idea of what a microscope should look like. The American firm of Charles Spenser began to produce world-beating objective lens systems in the early 1840s, heralding a final phase in the perfection of visual quality which has never been exceeded—indeed it is generally acknowledged that Victorian lenses are often superior to anything available today. This phenomenal improvement was in the main due to the development of 'apochromatic' lenses (the term means 'away from colour' as opposed to achromatic which can be defined as 'without colour') in which the colour correction was greater than ever before. By the use of a third refracting component in the lens system it was possible to obtain near-perfect colour correction.

The achromatic lens corrected for both red and blue light by bringing them to focus virtually at the same plane. The apochromat went further still by bringing the previously uncorrected region in the middle of the spectrum (the green rays) to focus. The inventor was Ernst Abbé, whose work gave Germany a pronounced lead over other European countries in the field. It

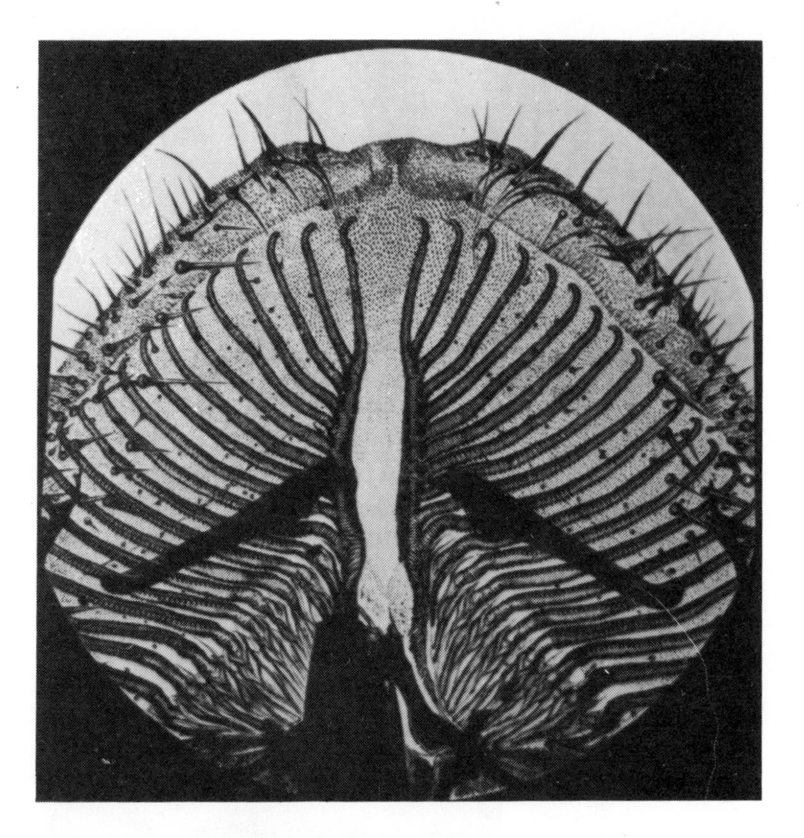

A century-old micrograph of a favourite Victorian test-object—the proboscis of a house-fly.

was he who perfected the oil immersion system which, in order to bring the objective lens as near as possible to the theoretical limits, utilizes a droplet of oil—itself acting as an additional lens element in some ways—which brought the objective into functional contact with the cover-slip of the object under study.

The way in which this occurs is simple to grasp if we bear in mind that the essential function of the microscope is to enable the eye to focus on objects at a shorter distance than is normally possible—and it is clear that, the nearer an object is to the eye, the larger it will appear. But as we have seen it is not enough to have merely high magnifications; unless the lens is able to reveal fine detail we will see nothing more, no matter how high the magnification itself may be. The effect of increasing the degree

Ultimate limits of optical microscopy—*c*.1870.
Markings on the diatom *Pleurosigma angulatum* photographed with an apochromatic objective of NA 1.30. Magnification × 3,000 approx. Micrograph by Dr R. Leitz.

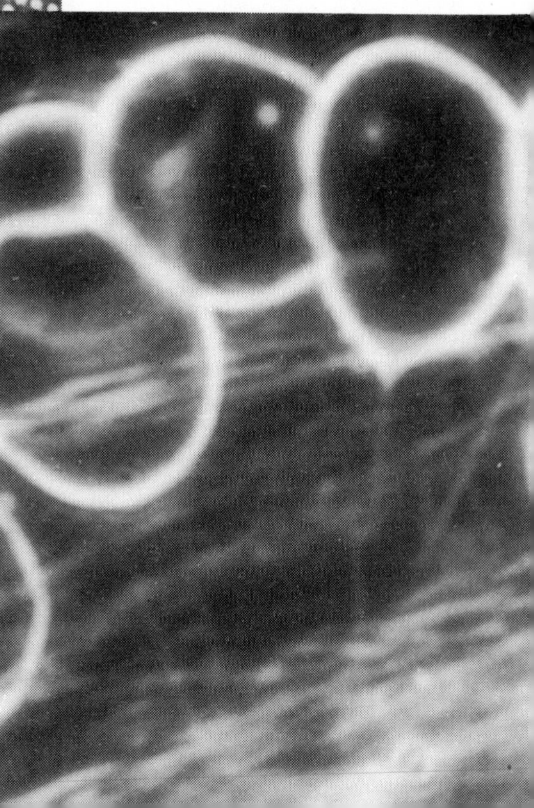

Ultimate limits of optical microscopy—*c*.1970.
Dark-ground view of coagulating human blood. The fine 'balloon-string' fibrils shown here for the first time are among the finest structures ever seen by the light microscope. Magnification × 3,000 approx. Micrograph by Brian J. Ford.

of magnification until the clarity of the object is lost is known as 'empty magnification'—and it is somewhat like attempting to see extra detail in a snapshot by putting a lens on it. The indistinctness is magnified as much as the object. The important factor is the lens's numerical aperture, abbreviated to NA. Mathematically the nature of NA is beyond the compass of this book, but it is in practice an index of the power of the lens to collect light and, by derivation, its powers of resolution. Theoretically perfect lenses of this type would have an NA of roughly 1.52. Typical oil-immersion lenses manufactured today are rated 1.30, which does not compare favourably with objective systems of 1.40 manufactured in Germany in the 1880s and others of 1.50 produced in England in the following decade. Ernst Abbé even designed one lens for the Zeiss manufacturing company which used a high-refraction immersion liquid (naphthalene monobromide) and was rated at an NA of 1.63. But only a few were ever made.

Of course, this is a technical matter—but it is important to realize that the numerical value of these Victorian lenses brings home one vital fact in the unravelling argument: before the turn of the last century, scientists had microscopes with an optical quality as good as, even better than, anything we have today. Thus, quietly and almost unannounced, the near-perfect microscope had arrived. What was needed now was the expertise to utilize it properly.

Man was now possessed of the way into the previously invisible universe. All he had to do now was explore, study—and attempt to understand.

5 The Science Comes of Age

BY THE turn of the century the groundwork had been done—and the real task of discovery and interpretation was about to begin. At last microscopy was exciting—and fashionistic too.

Intellectually, the discipline in its varied forms had established itself on a plateau; it was recognized that there were myriad discoveries to be made of a purely observational nature—microscopy was becoming an extended exercise in documentation. In some respects it is still so today. There are a good many papers published still which are records of pure observation—the structure of the penis of a cockroach, new types of rotifers with two knobs instead of four, and the microscopical configuration of elephant lymph glands are amongst actual recent examples. It was understood that living things conformed to the cell theory; that the overwhelming number of living things were composed of a single cell only, and that these micro-organisms abound in nature and may be isolated from mountain-tops, the bottom of the ocean, even hot sulphurous natural geysers.

And it had dawned on man that beneath his feet (and certainly beyond his ability to see without aid) was a teeming, active population of living things which were so diminutive that they could, in the space held by a salt-spoon, outnumber the human population of the earth—in fact they could exceed the total number of humans that have ever existed since the species *Homo sapiens* first appeared. Strangest of all, in some respects, was the revelation that higher living things, such as man himself, were composed of vast coherent communities of cells which in assembly produced the co-ordinated pattern of metabolism and activity that we call life. It was revealed that man's genius, his inventive curiosity, his ability to envisage, design, produce and—perhaps most marvellous of all—*control* motor cars, aircraft and the rest; his innate aesthetic abilities and his powers of creativity; even his lust, his instincts and his basic ability to live at all were the manifestations of a vast colony of separate living entities, so independent in their self-contained nature and yet so beautifully integrated into the cybernetic, regulated harmonies of the whole. It was, and is, a thesis of incomparable beauty, of inconceivable complexity and yet in essence so utterly, utterly simple.

But however simple the essential concept of the single cell undoubtedly was, it became plainly apparent that in structural and functional terms cells themselves were a complex and varied lot. Science was faced with some startling and difficult new notions as a result. The simple cell, much as we were taught at school, had been revealed as a body containing a nucleus (the controlling centre of the cell itself, bearing the essential hereditary matter) surrounded by and embedded in a jelly-like viscous fluid called protoplasm. Tiny bodies scattered around inside this protoplasmic system were given various names—Golgi apparatus, mitochondria—and have since turned out to be organelles of specific purpose (such as the control of the cell's machinery).

Such was—and is—the general picture of 'a typical cell'. But many cells are far from typical. Man's red blood cells (erythrocytes)—which were the first human cells to be observed, it will be recalled—are most exceptional in that they are anucleate—i.e., they contain no nucleus. But as if by contrast the most common kind of white blood corpuscle (the aptly named polymorphonuclear granulocyte) seems to contain several, as the large nucleus is lobed and divided into segments. Because the blood cells are so easily obtainable and since they float, detached, in a liquid matrix they have always been favourites for observation. Yet paradoxically they are very exceptional.

Animal cells show a vast variety of forms. The head of a single sperm, smaller by far than an erythrocyte, contains virtually all the genetic information to create a full-grown human organism in all its details, whilst at the other end of the scale we have the egg of an ostrich which is itself essentially a single cell. This interpretation of a bird's egg was first proposed by Gegenbaur in 1861 who wrote: 'They are nothing else than enormous cells peculiarly modified', and provoked considerable controversy when he did so. But he was correct. A bird's egg—unless it has been fertilized and contains a developing embryo—is essentially a gigantic cell, containing one nucleus and an accumulation of yolk to last the young bird through its period of imprisonment until it is mature enough to hatch.

In the plant world too there are similar extremes, from the minute blue-green algae which are too small to boast a properly organized nucleus up to the little-known *Caulerpaceae*, a family of deep-water seaweeds with large single-celled branches several feet in length. And these are only the extremes in the green plants. In the microscopic forms (including bacteria and fungi) even more minute varieties are found. So the plant cell may be only a little more than a thousandth of a millimetre across, or as much as a metre; clearly the diversity of the cell, as it became apparent, was staggering to the research workers. But the essential truth remained; the cell theory stood. It has been a mainstay of biological discovery and advance ever since.

By the end of the century the adoption of this view had greatly smoothed the unravelling of embryology, and it was in the other microscopical fields (such as histology, the study of tissues and cytology, the study of cells) that the firm foundation for all later work was laid down. Plant anatomy had been extensively investigated too, and the different structures in the growing plant had been examined, defined, interpreted; in short, the essential nature of living things had—in the space of a few decades—suddenly unravelled itself before an astonished and excited public for the first time.

Bacteria too were being cultivated in the laboratory. The culture techniques laid down in the late nineteenth century are essentially those we use today, and the development of the bacterial theory of disease was extended to plants in 1899 by the work of Erwin Frank Smith who solidly established that several parasitic diseases were due to bacteria, not fungi. Metallurgy had also begun to proliferate as a microscopic science, though the results were not clearly understood, and geologists were turning towards the minute structure of mineral specimens as a source of further study.

And the handling of materials had come of age: the standard microscope slide as we know it today had become commonplace. From the mid-seventeenth century objects held, in the main, on fine pins the technique had developed until—by the early eighteenth century—it was fashionable to mount objects between slivers of talc or mica held in ivory mounts. Glass plates, as we saw earlier, had anticipated the later use of slides.

The work of the histologists—like Virchow—had necessitated further precision, and the development of the improved optical systems already mentioned was a further factor which acted as a stimulus. Finally the new chemical industry had thrown up a range of dyes and stains by which previously transparent specimens could be coloured and better observed.

In short, though the scientist who died a century ago would have worked with techniques and apparatus which we would regard as primitive, by the year 1900 the microscopical laboratory had developed to such a state that today's research worker would have felt pretty much at home. The microscope had truly come of age, and the enormous task of discovery could begin. It had taken 250 years to build the technical environment necessary, to engender enthusiasm and unanimity of purpose in this fundamental field of science, and to come to grips with the basic principles involved. Now man could begin to know the enemy, even if, as yet, he could not do much about it.

The greatest single problem faced by these research scientists was that of disease. It was now certain that a great many diseases (including all of the truly infectious conditions of man) were caused by micro-organisms of one sort or another—the problem was to find a cure. During the nineteenth century arsenic and mercury were used in the belief that they could poison the disease without killing the man. Poultices of hot bread or mustard were widely used, and anointing with solutions of iodine or carbolic acid was variously recommended. Often the essential dictum was: 'if the tincture is too strong it will bring off the skin, if too weak it will not hurt, but neither will it be efficacious.' For the unfortunate invalid it was a disturbing and painful prospect.

This new microbial concept threw all these old-fashioned ideas on the nature of disease to the wind. It was understood that 'germs' as such caused infection, and it was therefore necessary to kill them off in order to cure the condition—a simple and obvious doctrine. But how? In 1900 a biologist named Laveran (who had discovered the causative organism of malaria) turned his attention to the trypanosome, a small, snake-like flattened organism which infects many animals (one species of which also causes sleeping-sickness in man). He discovered that one of the species found in horses would grow satisfactorily in mice, giving him an excellent experimental subject for his trials. His results, published a year later, described how he treated these mice with arsenic, and had observed the trypanosomes to diminish in number in the blood of the animals—but still they died, sooner or later. It was not a cure, but it was demonstrably a help. A step in the right direction, at least, had been taken.

The paper was read by Paul Ehrlich, an earnest and enthusiastic scientist working at Frankfurt-am-Main, and to him it acted as a catalyst—it triggered off a chain reaction of ideas which were destined to make profound inroads into our approach to the cure of disease. He saw the mice as a perfect experimental animal, and the trypanosomes as an ideal form of 'germ': they killed 100 per cent of the mice they infected, yet they could not harm man. It was an ideal set-up for a research programme. And it reminded him of an earlier thesis he had advanced—a concept of great simplicity, and of considerable import. He had observed the staining of bacterial slides with interest. The stains coloured the bacteria vividly, yet did not stain everything with the same intensity—some tissue constituents were left largely unstained. Similarly, he observed that the dye methylene blue (a common laboratory stain) had an affinity for the granules within nerve cells, but did not stain other tissues so readily. And he had wondered whether a drug could be found which acted in this way. Obviously, the selective activities of these dyes showed that some chemicals could be selected to exert an effect on specific types of cell; and he felt that if only the dye could be specific for bacteria, and if it could additionally have a destructive effect on them, then clearly he had the makings of a successful anti-bacterial drug. Surely, he reasoned, these dyes could not attach themselves so very strongly to these vital structures without damaging them chemically—so he set out to test a vast range of dyes in the hope of finding one which exerted the desired effect.

It was a tortuous and far-fetched programme of research which he set himself. Obsessed with the essential self-evident truth of the theory he was determined to find the chemical which would perform this miracle. In all he tested over five hundred chemical dyes on mice injected with trypanosomes and each and every one failed him. The nearest he came to success during this trying period was with a dye later named Trypan Red. It did, for the first time, cure some of his experimental mice. But it did nothing for the majority, and was quite useless when he repeated the experiments with other species of animal.

None the less, it was the first whiff of the success that was to come. His next important observations were made on a drug named atoxyl, which turned more people blind than it had ever cured; it was a molecule based on the structure of benzene—six atoms of carbon in a ring—with arsenic added on in place of one of the side-groups. He and his chemist friends worked on the molecule, changing it, altering its structure subtly, and always testing, testing, testing. At this point Ehrlich wrote, 'we shall prepare a thousand or even more compounds.'

But it was not to be necessary after all. The 592nd compound he tested showed a sudden, dramatic improvement. Its effectiveness was increased, and it was not particularly toxic to the mice. Excited, he pressed on until the time came to test the 606th chemical on the list. Its toxicity was low, yet in mice it killed the trypanosomes with ease. In fact, as field trials proved, it was not very applicable to trypanosomiasis in man—but continued research suddenly revealed an application which was to have world-wide repercussions. For compound 606 was lethal to *Treponema pallidum*, the causative organism of syphilis. It was in September 1909 that the first tests on an experimental rabbit demonstrated the efficacy of the cure—the syphilitic rabbit was made free of the organisms after only one injection. It was a moment of medical history, and Ehrlich, working in his crowded, dusty study surrounded by literally hundredweights of journals, textbooks and encyclopaedias, knew then that his long search had ended. 'After years of misfortune and failure I had one moment of luck,' he said later.

The compound was named *salvarsan*. And it heralded a new science—chemotherapy. It was to do more than any other single discipline to reduce the suffering of diseased and dying men the world over. More important still, it had the effect of encouraging others to look for similar chemotherapeutic agents—and the sulpha drugs were the direct result of this stimulus. Eventually it would lead to the discovery of antibiotics, too, though that is a subject to which we will return in due course.

But all the time that the discovery of salvarsan was making the conquest of microbes seem nearer, there was a quite different field of investigation which began to turn in the opposite direction. It had started, in some respects, with the Rembrandt tulips—those characteristically red and yellow blooms so highly prized in the Renaissance—which were very difficult to breed successfully. Even in the pre-genetics era it was difficult for plant breeders to see how it was possible for tulips to take on such a strange and apparently random display of colour, in spite of the care and devotion with which the parent strains were chosen. They were not to know it, of course, but they were witnessing an *infection* of the tulips.

The first records of any systematic investigation into such a plant disease are those of Adolf Meyer, a German biologist who made a study of tobacco mosaic disease—a condition in which the leaves of the plants become mottled and weakened: it was a condition of considerable economic importance. He found that the disease was transmissible through the sap extracted from the leaves of infected plants. But he could not find any bacterium which seemed to be responsible. Oddly enough, in 1886 or thereabouts when he was carrying out his experiments, Pasteur was busy in his studies of rabies—and he too failed to find a bacterium in his samples of infective serum.

However, Meyer went enthusiastically on his way, examining sap samples for the micro-organism responsible and inoculating healthy plants with samples rich in bacteria (such as pigeon-droppings and mouldy, aged cheese) in the hope of stumbling across the culprit, but to no avail.

It was a Russian, Ivanovski, who took up the studies in the next few years and in 1892 published some interesting conclusions. He agreed with Meyer in that the infective principle was found in sap, and that boiling (or even near-boiling) inactivated it. But he went further, and showed that the crystal-clear filtrate which emerged when infectious samples were passed under pressure through a porcelain filter (which removes all bacteria) was still able to carry the disease—he wrote that he was probably dealing with a very minute form of bacteria. But that, though a distinct improvement on Meyer's thesis (which had stated, for some reason, that the infective principle was removed by passing the sap through ordinary filter-paper), was still not quite accurate.

It was a Dutchman, M. W. Beijerinck, who put the matter into its conceptual context by recognizing that any infective principle on such a diminutive scale and with such strange propensities could not ordinarily be a bacterium at all. He conceived the phrase *contagium vivum fluidum*—a long-winded and verbose expression, certainly; but it was the first hint that there was some sub-bacterial disease agent. In his writing he described the principle as a 'virus'—the term we use today.

Two German bacteriologists, Frosch and Dahlem, found in 1900 that foot-and-mouth disease was also due to a 'filterable virus' and so the new term began to acquire some currency. However it was a gradual process by which the term gained its present-day meaning. The concept of a 'filterable virus' was then taken to mean something with the quality of infectivity about it, but which was too small to be filtered out of solution in the normal way. Now it means a quite specific type of entity (with some border-line cases which keep the picture still confused) with certain definite characteristics. But in the early years of the century the term became a convenient repository for any disease which seemed to have an unidentifiable cause. The vagueness is apparent when we look back at Beijerinck's comments on one of Frosch's co-workers, Löffler; the Dutchman was convinced that he was dealing with a soluble, infective protein of some kind whereas Löffler (whose views Beijerinck decried) maintained that the infective agent was a minute, but particulate, body. They were, as it happens, both right.

The effect of this debate over the nature of the 'filterable virus' added a stimulus to those who were seeking greater and more detailed powers of magnification—increased resolution, in other words. Many microscopists (particularly Van Heurck) used deep blue light for their micrographs in the latter part of the nineteenth century since resolution—being a function of the wavelength of light—was markedly greater when these short-wave blue light sources were used. But the greatest increase would derive from the use of ultra-violet rays with only half the wavelength of ordinary light sources: twice the resolving power was possible, in theory at least. But there are difficulties. In the first place ultra-violet rays are invisible to the human eye, and so the image had to be viewed on a fluorescent screen if it was to be seen at all—direct viewing of the object was impossible. And secondly, ordinary glass lenses are in general opaque to ultra-violet, so none of the illuminating beam would reach the specimen anyway—a formidable pair of practical drawbacks.

Two workers at the Zeiss works in Germany, inspired by the calculations of Abbé, set about making an ultra-violet microscope in 1904. They were Köhler and Von Rohr, who made the lenses from fused quartz (since this is virtually transparent to the ultra-violet beam)—and the slides and cover-slips they utilized were fashioned from the same mineral. In fact, the facility of resolving details half the size of those that could be previously seen conferred only marginal advantages. There are only one or two types of specimen (such as the virus of smallpox and that of psittacosis, both of which are at the limits of optical resolution) which are in the size range that can benefit from the technique. What with this drawback, and the cumbersome and time-consuming methods of operation (not to mention the expense involved in the special manufacture of lenses, etc.), the system was never widely used.

But it did throw up one useful new branch of the science which is important in today's research scene—the discipline of fluorescence microscopy. Ultra-violet wavelengths make many chemical compounds glow in the dark, and it has been possible to observe the activities of antibodies—the body's main system of defence against infection—by 'labelling' them with a fluorescent dye and then by observing the preparation under a fluorescence microscope. Ultra-violet is used as the illuminant, but its effect is to make the labelled compounds glow brightly. In this way the pathways can be clearly followed, and the fact that it is ordinary visible light which is observed (emitted by the glowing fluorescent dyes) eliminates the need for objective and eyepiece lenses made of special materials. This is a specialized form of modern microscopy which we need not consider closely—but which does stem directly from the aims of Köhler and Von Rohr at the commencement of this century.

Other workers, taking the broad view of the prospects opening up in microbiology, realized the considerable potential for research on micro-organisms *in vitro*—literally, in glassware—and initiated far-reaching physiological experiments. One of these was a Frenchman, E. Wildiers, who carried out some detailed studies of the nutritional behaviour of yeast. To do this, he took some of the synthetic growth medium advocated by Pasteur—a solution of inorganic materials and sugar, without any other organic residues. Pasteur had found that yeasts could grow in such a solution. But Wildiers, when he repeated the same experiments, disagreed—he claimed that the yeasts would not proliferate at all, or if they did it was very slowly. Where was the discrepancy? It was simple—Pasteur had added quite sizeable amounts of yeast cells to his cultures, whereas Wildiers added only a few cells to start the colony growing. There must be something, argued Wildiers, in the yeast cells which encourages them to grow: in small inocula containing only a few cells there is not enough of this necessary material. All very well, but how to prove it?

This experiment was not difficult. He carried out several trials. First he added just a few cells to the medium and, by contrast, set up similar cultures in which he had placed larger numbers of cells. The latter cultures grew and proliferated, whereas those with a small number did not grow at all. Then he boiled some yeast and added it to a series of flasks before inoculation with live yeast; and he found that these all grew well, whereas cultures without the boiled additive did not. So far so good—he had proved the flaw in Pasteur's argument, and had demonstrated the need for boiled yeast (or some extract of it) if living yeast cells were to proliferate. But was the essential principle carried in the yeast cells themselves, or was it in the solution in which they had been boiled? That was simple: he divided the boiled samples in two—cells in one, supernatant fluid in the other—and added these to culture media. This time he saw that the growth was heavy in the medium containing the liquid additive extracted from the yeast cells, but in the cultures containing the boiled yeast cells alone there was no growth.

And so, he deduced, there was a growth substance present in aqueous extracts of yeast cells. He called it 'bios'. Some years later Funk extended the work, and in 1912 he felt (erroneously, as it happens) that the mystery chemical was an amine—one of a group of biochemical compounds. And since it was an amine which was necessary for the vital processes of a cell, he christened it 'vitamine'. From this simple experimental basis has since grown the science of nutrition, which has evolved in this century from an underrated sideline into a vast new discipline: and deficiency diseases—rickets, scurvy, beri-beri and the rest—have now been better understood and, at least in the developed countries, controlled. So an important part of public health arose from some simple microscopical observations on the humble yeast cell.

The microscope was revealing the way to an understanding of mankind in other ways, too. It was showing how adults developed from a single egg cell into the intricate, majestic complexity of mature manhood. All of this work was carried out with the aid of the microscope—much of it being done on plants and animals and related only later to man.

Haeckel in 1877 had written a long account of embryology in which he sketched the overall changes of vertebrate development from a single cell that was, he believed, anucleate (i.e., without a nucleus). In that respect he was wrong—but his basic overall treatise was sound enough. In 1880 Balfour's work appeared, and in 1886 Hertwig's accounts helped to complete the foundation of modern embryological theory. By the early part of the present century tissue culture (the growth of living tissues in glass containers outside the body, generally in some artificial nutrient solution) had been perfected and many research workers grew cells in warm chambers under the microscope, searching for the way in which man came to reach his state of sophisticated complexity. In this way the development of man, through the revelations of the microscope, was systematically unravelled—except for one fundamental aspect of the matter.

It was this: why did a man develop from the human ovum at all? Why was it impossible for an elephant to give birth to a ram, or an orchid to spawn daffodils? It is not such a foolish question as it seems.

An ovum (whether the egg-cell of a human or a lowly plant species) is only a living cell, after all; there is nothing very 'humanoid' about the human egg. It is just a collection of cytoplasm with its various microscopic organelles parcelled neatly around a nucleus. The human ovum is the largest cell in the body (it is, as it happens, the only human cell large enough to be just visible to the unaided eye) but there was no reason whatever to show why it had to grow into a new human being. There was nothing that could alter this fate, short of destruction, and it was clear to the philosopher scientists at the turn of the century that there was somewhere a complex, hidden network of information —information that could set this single cell to work, collecting and using foodstuffs and assembling inconceivably intricate arrays of chemical molecules into a functioning, intelligent wide-awake man.

But what? And where was it? No-one seemed to know: everyone was content, instead, to assume that it was somehow preordained for human cells to grow into human babies, and for buttercups to proliferate their own species yet no other. All the same, there had to be a reason. And it was, surely, somewhere in that mysterious, amorphous blob at the heart of most cells—the nucleus. It was accepted that this was the 'brain' of the cell. But though this notion helped to locate the centre of genetic control, it did nothing at all to unravel its mechanisms, nor was it helpful in indicating—even in outline—wherein lay the secret.

Back in 1830, Amici, in a brilliant and delicate programme of research, had followed the structure of the pollen tube as it grew from the pollen grains down, through a flower's pistil into the ovule; he had observed fertilization under the microscope. For forty years or so the controversy which this initiated raged back and forth across Europe, but eventually it was generally agreed that the pollen grains were the male elements of reproduction, and the ovule the female; in 1855 a microscopist named Pringsheim watched free-swimming algal spermatozooids enter the female egg-cell and twenty-four years later a Swiss scientist, Fol, watched the same phenomenon in animal cells under his lens. At about the same time Strasburger had watched the nucleus of the male and female gametes fuse into one in plant specimens, and Fol confirmed this occurrence in the animal kingdom.

At once the German Weismann saw one important implication of this. If nuclei from the father and mother fused together to make the nucleus of the offspring cell, there would be a doubling of the amount of genetic material. This doubling would take place with each generation, so that all offspring would have twice as much genetic material as their parents—a ludicrous situation. Weismann postulated that, somewhere prior to the fusion of male and female cells, there must be a halving of the amount of genetic matter so that the half-quota in the two sex cells would restore the normal state in the ovum.

Van Benenden at once stepped in to name the normal adult cells as 'diploid', and the germ cells (i.e., the sex cells—the term 'germ' is used here as a direct result of the earlier concept of *seminaria*, *q.v.*) he called 'haploid'. He had made one further important discovery—many workers before him had noticed the formation of chromosomes in the centre of dividing cells. The chromosomes were ribbon-like structures (sometimes rod-shaped) which appeared to condense out of the dividing nucleus. They split into two parts, re-formed nuclei, and the cell structure was then divided in two. That, it was known, was how cells multiplied. But Van Benenden took it further. He noticed that the number of chromosomes in each mature cell of any species was the same as in all the others. But, he observed, the reproductive cells have only half the number in the somatic (adult) cells. This, surely, was the key to Weismann's 'doubling-up' theory.

Within the next few years leading up to the turn of the century, the subject halted for a while as if to gain breath. Then, in 1900, de Vries began a survey of earlier literature in search of experimental evidence which could confirm (or deny) the existence of genetic units that acted in the same way as chromosomes were now observed to behave. He could, quite easily, have launched some selective breeding experiments himself; but they take time, and de Vries did not want to waste any. So he looked back at the results of others—and he found one person* who had had plenty of time for experimentation.

That man was Gregor Mendel, a monk of Brno in Czechoslovakia, who—almost as a 'professional necessity'—had an abundance of time for reflection and thought. De Vries, when he located Mendel's writings on heredity, saw how he had bred sweet peas. He had crossed tall and short varieties and found that, instead of the medium-height offspring that he would have expected, the seeds produced tall and short plants in the ratio of 3:1 respectively. It would be pointless to reiterate here all his work, and students of school biology will be familar with the 'Punnet-square' analysis which enables Mendelian predictions to

*In this case, too, there were precedents. But Mendel did more than anyone else to found the science of inheritance.

be carried out. But even these first results had a clear message: hereditable characteristics do not simply 'blend', as the popular teachings of Darwinism implied. They were either present, or absent, or they might be masked; but simple genetic blending did not occur.*

Mendel's genetic laws were clear and intelligently simple; they were the result of a mathematical interpretation of his observations on the cross-bred sweet pea plants. De Vries, and his contemporaries, found them valuable evidence in favour of particulate inheritance, as we may call it; and in 1902 an American, Sutton, put his finger on an important truth. Once again it was not the startling breakthrough of an immense intelligence, merely the coining of something that was—by then—almost self-evident; but still it had to be said. Sutton described how the division of chromosomes, and in particular the formation of the reproductive cells (in which the chromosome number is halved), exactly paralleled Mendel's principles.

Surely, he argued, the correlation was more than coincidence: and he firmly argued that Mendel's 'genetic characteristics' were part and parcel of the chromosomes. These were the bodies that carried hereditable matter, he claimed; and in observing the chromosomes we were watching the very source of man's nature.

In 1906 experiments were begun on *Drosophila*, the fruit fly, which is ideally adapted for genetic experiments. It has but four chromosomes, a short generation time and a rapid breeding rate; and a very large number of genetic experiments have been since carried out until we can with confidence claim to know our way around the chromosomes of *Drosophila*, pointing our fingers at the exact source of the red-eye gene, or the complex that stunts wing development. This research has paved the way for a greater understanding of man's inheritance.

And so in the early part of this century, most of our 'new sciences' in biology and medicine had been founded. Clearly they owed much to the microscope, and some owed all. Yet many

*At that time there was a 'Law of Ancestral Inheritance' which suggested that the genetic constitution of any individual was one-half due to his parents, one-quarter to his grandparents, one-eighth to his great-grandparents and so on. There was, according to this thesis, a little bit of Adam in us all.

important new discoveries were still tantalizingly distant—and the discovery of antibiotics is perhaps the most significant single example.

Whilst Ehrlich, in a comparative blaze of publicity, had launched salvarsan and so founded the discipline of chemotherapy, other microbiologists were unobtrusively tackling the same problem from an entirely different angle. They were investigating phenomena where one living species seemed to antagonize the growth and development of another. In 1876 Tyndall, the physicist, had noticed that a growth of fungus formed on some broth cultures in which bacteria were growing.* More than that, he observed that the bacteria were apparently being destroyed—or at least prevented from growing further—by the fungus. He concluded that it must be due to competition for oxygen that led to the death of the bacteria; the felty mass of the fungus, he recorded, seemed to block off fresh supplies of oxygen from reaching the bacteria, with the result that they died.

There are questions one would raise. Why did not other forms of bacteria less needful of oxygen proliferate in their place, for instance? And what was the identity of the fungus? Under the microscope it showed a quite characteristic means of spore formation; the slender thread-like hyphae of the fungus divided a few times to form a brush-like structure. Then, each single fibre of the 'brush' divided up into a chain of hardened, resistant spores. So characteristic was the appearance under the microscope that the fungus had even been named after the brush it so closely resembled. The Latin for a brush is *penicillus*; the fungus was *Penicillium*—and was that the first recorded observation of penicillin in action?

Though one can be sure only about our inability ever to solve that mystery, it is, I am sure, certainly very possible that Tyndall was observing antibiotics at work. Nothing came of it, however. Science, and the attitudes of scientists, were not yet attuned to

*Tyndall gave his name to a little-used process by which broth culture-fluids or other liquid substances—including foodstuffs—may be rendered free of organisms. The procedure is to boil the liquid for twenty minutes on three consecutive days, whilst excluding any fresh contamination. During the 'cool' periods, he argued, any bacterial spores that survive will germinate and the next heating session will then kill the vulnerable young cells. 'Tyndallization', as it is known, is occasionally used to sterilize heat-sensitive liquids.

the possibilities. The fact that antagonism of this sort existed, of course, was evident to any microbiologist; and the possibility that some therapeutic agent of use to man was a fairly straightforward consequence of the observation. Pasteur wrote in the following year, 1877, that the antagonism he had observed between 'common bacteria' and the organism of anthrax 'perhaps justify the highest hopes for therapeutics'. And the Italian Cantani, in 1885, claimed to have observed the inhibition of tuberculosis by the antagonism of bacteria he described as *Bact. termo* (an unidentifiable species, and probably a mixture of several common bacillary organisms). He wrote a paper with the singularly far-sighted title *Un tentativo di bacterioterapia*—'possibilities for bacterial therapy'—in that year.

By now the antagonistic propensities of certain species were well known, and in 1889 the term 'antibiotic'—literally, *against life*, but implying *life-against-life* in accepted usage—was coined.

In 1896 a bacteriologist named Gosio found that a mould growing on spoiled maize seemed to have an inhibitory effect on some bacteria. The fungus was *Penicillium brevis-compactum*, and he crystallized an extract with pronounced antibiotic properties which would inhibit the organism of anthrax. The extract was not penicillin; but it was certainly a step towards the as yet unseen goal.

In 1903 Lode discovered a bacterial extract that could inhibit the growth of anthrax and *Staphylococcus aureus* (a common organism causing conditions ranging from boils to septicaemia). He found that it was not an enzyme, that the chemical was destroyed by heat, but that it could be isolated by distillation in a vacuum; it was soluble in alcohol, he found, though not in ether. His research was immensely careful and clear; but the extract did not work against infections experimentally induced in mice.

Meanwhile an extract of the bacterium *Pseudomonas pyocyanea* had been used to help combat the effects of diphtheria. The extract was known as *pyocyanase* and it was generally squirted into the throats of sufferers. Certainly it has the ring of 'quack' medicine about it, but there was evidence that it helped more than it hindered—and though the technique petered out until it was gone by the end of the First World War, it was certainly another stab at the target. Also during the early years of the century, the concept of 'replacement therapy' was born—and died. It taught that injections of certain bacteria would overcome and destroy those that were causing the patient's illness. Both of these practices arose from several concurrent sources, and they reflected the balance of opinion at that time. For several years it was widely held, for example, that it was beneficial to give a patient erysipelas since it would cure any pre-existing infection.

Though the fashion for 'replacement therapy' soon died, the scientific evidence of antagonism was accumulating. Between 1910 and 1913 Vaudremer used an extract of *Aspergillus fumigatus* (a 'cousin' of *Penicillium*) in the treatment of over two hundred tubercular patients. The results, he wrote, were generally equivocal—but in some cases he was sure the extract had helped. Even earlier, in 1904, Tartakovskii had made a similar extract of *Penicillium glaucum* and had found it to inhibit the growth of the intestinal bacterium *Escherichia coli*. And in 1913 the British worker Black and his colleagues isolated *penicillic acid* from the fungus *Penicillium*—Sir Howard Florey, who played such an important role in the later development of antibiotics, once told me that this substance was truly to be considered an antibiotic. All this—and the First World War was yet to start.

Between the war years and the mid 1920s a whole range of fungi were found to produce inhibitory or antibiotic substances. Apart from *Penicillium* and *Aspergillus* which we have already considered there were *Lentinus*, *Spicaria*, *Fusaria*, *Sparassis*, *Cladosporium*. And all the time, in London, a bacteriologist named Fleming was busy studying culture plates of *Staphylococcus* in the laboratory of St Mary's Hospital, Paddington.

His observations necessitated the inspection of culture plates under a binocular microscope; to do this he removed the lid of the petri dish in which the organisms had grown and inevitably the contents was liable to be contaminated by air-borne spores of one kind or another. One day in 1928 Fleming (or someone else in his laboratory—at least one suggestion was made that the phenomenon was first noticed by a Jewish laboratory assistant) saw that one of the culture plates was contaminated by a fungus colony. Most strange of all, the bacterial colonies immediately adjacent to the colony were undergoing lysis—they were literally dissolving away.

What happened since is history, of course, and there are many detailed accounts of the development of penicillin. So we cannot go into that subject here—in any event, it soon outgrew its microscopical beginnings, like so many of the discoveries we have described, and entered the realm of process biochemistry.

But the early work, Fleming's first published research on the subject, was centred on microbiology. His paper (which appeared in 1929) discussed the properties of the fungus in general terms, and it now makes interesting reading. For instance in his book *Penicillin*, published in 1946, Fleming states that he gave the name to the 'antibacterial substance' produced by the fungus, in the same way as previous drugs such as digitalin had been named after *Digitalis* (the foxglove) and aloin from aloes. But Fleming's statement in this book is incorrect. For in his original paper it can be seen that he applied the name 'penicillin' *not* to the drug at all—but to the broth medium in which the fungus had been cultured. He applied it to the 'mould broth filtrate', to use his own words, and not to the antibiotic it contained. This is, in scientific terms, an important distinction. It is as though he had identified wine as alcohol, instead of merely *containing* it—and had later written as though he had never said it at all.

At any rate, it was somewhat naïve to give the name to the 'filtrate'. Had he followed scientific precedent, he would have stated that the antibacterial substance was yet to be isolated.

The precedents he cited were not followed so exactly, after all; digitalin is the crystalline alkaloid extracted from foxglove leaves —not merely the crude sap.

Similarly, though Fleming's later writings tend to give the impression that even in 1929 he realized what a powerful drug he had discovered, this first paper shows only the vaguest outlines of how it might be used. He was well aware of penicillin's instability, and the main use to which he suggests it should be put was the isolation of bacteria in the laboratory.

The unravelling of the nature of the drug in biochemical terms was the work of Dr H. Raistrick of the London School of Hygiene, who published his first findings in the *Biochemical Journal* in 1932. Not only did he carry out the first detailed chemical examination of penicillin—the compound, that is, not the broth—and recognize that it was a specific entity (as Fleming had not done), but he tried to initiate some clinical trials of the drug as a therapeutic agent. But the medical world was not ready. Chemotherapy was still relatively new; and the unfashionistic idea of using a fungus extract was regarded as somewhat 'lunatic fringe' at that time. As one of Raistrick's colleagues wrote later, 'his medical colleagues would not listen to his pleadings, and he could get no clinical tests made'. Raistrick, it is clear, knew what he was on to. But the world of medicine was not yet ready; yet again, the seed of discovery fell on unprepared, inhospitable ground.

In 1935, three years after Raistrick published his findings, Gerhard Domagk, a German biochemist, published a short paper entitled '*Ein Betrag zur Chemotherapie der bakteriellen Infektionen*'—a contribution to the chemotherapy of bacterial infections—which really had nothing to do with penicillin, but which started the ripples spreading. In this paper he described prontosil, the first of the sulpha-drugs. It was interesting because it was the continuation of Ehrlich's teachings—to test varied biochemical compounds, no matter how unlikely they looked—and it was all based on animal experiments. Domagk noted that the drug had no effect on bacteria *in vitro*. Cultures subjected to it thrived as actively as they had done before. But in the animal its effect was salutary. It wiped out streptococcal and staphylococcal infections rapidly—better than anything that had been seen before. The reason for this was uncovered during the following years' research: prontosil was changed by the body into a different, allied compound—sulphanilamide.

This discovery was in itself important, of course; but it had two other effects. The first was to awaken medical opinion to the existence of powerful chemical antibacterial substances that could be used in man to fight disease—with the corollary that the activity of the product might be due to vital, living processes and not merely to basic biochemistry—and the second was to make British scientists realize that German pharmacists must have had a very powerful range of chemotherapeutic agents available by then (Domagk's work in 1935 had indicated how far they had progressed).

Fleming wrote in 1946 that in the 1930s a 'remarkable change has happened in medical thought in regard to the chemotherapy of bacterial infections . . . there was an idea that the common pyogenic cocci, after they invaded the body, were beyond the reach of chemicals'. But after prontosil had been discovered, he stated, 'the medical profession became chemotherapeutically minded'. This interesting comment demonstrates graphically how assumption and 'instinctive unanimity', as we may call it, can persistently block progress. Between the time when Raistrick was 'pleading' and Florey was looking back over the literature there was a fundamental change of attitude which made chemotherapy seem, somehow, more popular—fashionistic, almost. There was not a single scrap of extra experimental evidence to change opinions about penicillin during all this time; it simply became a bandwagon. Those who had scoffed in 1932 were enthusing eight years later—but without any change whatever in the facts of the matter.

Here again we may observe the tragedy of human nature in its attitude to science. We can see the importance of fashionism—and the secondary, almost minor significance of accuracy, objectivity, and insight in the prosecution of a new discovery. There was no lack of scientific evidence in favour of penicillin. There was ample enthusiasm from several quarters (and the almost fanatical campaigning by Raistrick) in favour of clinical investigation—indeed with the best will in the world it is impossible to see why work on penicillin did not start ten years earlier. Until, that is, we take into account one of the inevitable conclusions of this book: that the mind of the recipient of any idea must be ready, or it will simply not accept it. No matter how clearly self-evident it may be, its greatest obstacle is inertia, stubbornness, the academic herd-instinct. Whether it is a new washing-powder, a broom from the suitcase of a door-to-door hawker, or a life-saving drug of enormous potential—it has to be *sold.*

Once again there was a situation in which the efforts of the pioneers had been forgotten and overlooked, but which had left the legacy of greater preparedness in the minds of men. And so it was at the turn of the decade that Howard Florey and a team of co-workers decided to take a more serious retrospective look at the early work on antibiosis. They found, as we have seen, many precedents. And—best of all—when they wrote to Fleming they discovered that he still had a culture of *Penicillium* in his laboratory.

It was in this way that the significance of the antibiotic

principle first emerged. Small amounts of the drug were produced for tests, which were encouraging. British financial and technological support for the production of the drug were not readily forthcoming, and as the world knows it was the USA which began the mass-production of the drug on a production-line basis. And here was one of those fortunate chance discoveries with which scientific progress will always be blessed—a discovery without which penicillin might have faced formidable obstacles in the path of mass production.

Whilst experiments were under way in the USA during the early war years—experiments designed to find the best way of producing commercial amounts of the antibiotic—many different growth media were used. Like most fungi, the mould from which penicillin was extracted* could be grown in a wide range of organic culture solutions. One of those that was tried was corn steep liquor (a waste product of the maize processing industry). Initial tests showed that *Penicillium* grew satisfactorily in media containing this liquid. Not only that, but the rate of growth was faster than was customarily seen—and the production of penicillin far exceeded any earlier trials. It was obvious that there were mysterious, unidentifiable factors in the liquor which stimulated the production of large amounts of the antibiotic by the fungus. No-one knew how it worked, or why; but that was hardly important. What mattered was that the drug could be produced in larger amounts than had previously seemed possible.

So penicillin came to the aid of the soldier in the Allied war effort—all because of long-forgotten pioneers who had laid down the foundations of chemotherapy at the beginning of the century, the fact that Fleming had kept his cultures of *Penicillium* alive as a laboratory curiosity, and because someone wondered whether corn steep liquor might make a culture medium after all.

*In Fleming's original paper the fungus had been identified as *Penicillium rubrum*, but when Raistrick and his colleagues got to grips with the problem (and as we have seen, did most of the ground-work) they soon found that this was not its correct identity. In fact the species proved to be *P. notatum*. It had been first discovered and named by a scientist from Uppsala University called Westling, who found it in 1911. There was nothing portentous about it then. When Westling found the fungus it was growing on a pile of decaying hyssop—a fragrant herb.

The rest of the story is history, of course; and the legacy of the antibiotics (in terms of millions of lives saved each year) is common knowledge. Many of the bacterial diseases from which our forefathers suffered and died are now widely controlled: anthrax, plague, cholera, diphtheria, dysentery, many forms of meningitis and pneumonia, to select some of the more prominent examples. Screening many forms of fungi has thrown up a whole range of alternative antibiotics to choose from—and abuse has emerged too. Over-prescription of antibiotics as 'precautionary' drugs (prophylactics) rather than to cure a specific illness was seen a few years back, and until quite recently there were disturbing accounts of their over-usage in animal husbandry. As we have seen from accounts in the press as well as in the scientific literature, many species of bacteria have begun to show resistance to antibiotics which formerly killed them. So the need for new discoveries is not ended.

The further development of antibiotics—like the exploitation of vaccines—relied more on the mass-production techniques of the technologist than it did on the laboratory worker with a microscope to hand. In the late 1950s, indeed, it was found possible to alter the penicillin molecule by purely chemical means and so create new antibiotics with improved potency. For research of this kind a microscope was totally superfluous.

But if the microscope was being outgrown by the discipline in this field, there was another aspect of microscopy in which it had yet to score. This was in the study of viruses, those mysterious border-line half-living, half-inanimate entities which we left on page 121.

Research soon showed that the term by which they were then generally known, 'filterable virus', was not universally applicable. As if to emphasize the point, it was found that the virus particles of the first two such diseases to qualify for research in the public eye—namely smallpox and rabies—would not pass through the porcelain filters at all. And so the 'filterable' was dropped, and the term 'virus' became attached to a more definite concept: a class of organisms, almost.

What were they? Over the years that followed it was found that viruses contain nucleo-protein (hereditary material, DNA or RNA, of the sort found in all living cells), they were infectious and often disease-causing (pathogenic). They enter the cells of the host and therein undergo proliferation. These are clearly the characteristics of a living organism.

However there were many other qualities of the typical virus which shed quite a different light on the matter. They are, for a start, unable to grow, unable to divide as cells do; they do not produce normal enzymes—the chemicals responsible for the chemical reactions of life—and indeed it was found that they could be crystallized from solution, exactly like a chemical compound.

The debate as to whether viruses are living or not has continued with unabashed enthusiasm over the years, and is still current. Some insist that any parasitic nucleo-protein entity with the capacity to reproduce is obviously living, and others retort that a crystallizable substance with none of the qualities of metabolism is clearly not. The controversy is active, as most scientific controversies are, not merely as an academic question of definitions, but in personal, almost vindictive terms.

Between the wars there were many, many scientists who turned their attention to the virus—too many by far for us to enumerate. They used a variety of methods to obtain their results. Most of these investigators tackled the virus *en masse*, analysing it, culturing it, harvesting it, filtering and purifying it, crystallizing it; laboratory culture techniques showed that viruses (due to their properties described above) could not reproduce outside the living cell. The virus seemed to act like some kind of 'infectious gene', which simply took over the cell and instructed the host cell to manufacture more virus particles. This is a unique means of reproduction; the term 'replication' was coined to describe it. Through filtration and several other techniques a good idea of the size of virus particles was gained. And the use of standard medical procedures led in turn to the production of vaccines against some virus diseases.

Yet there were two important loopholes in the argument. First, there was no cure for any virus disease (and this is, as we shall see, virtually as true today as it was then). Man was unable to do anything after a virus had struck—other than sweat it out. Vaccines would make one immune, perhaps; but nothing could make you well once the disease had taken hold. Virus diseases (such as influenza) are still the greatest epidemiological headache faced by man today. The end of the First World War was marked by an international outbreak of flu which served to remind one of its terror. In Britain alone 150,000 people died of the disease; elsewhere the white population of the globe was showing a death-rate of around five per thousand whilst in India it was far higher. Five million people died in that country alone during the outbreak. And of course, since there is still no vaccine which is universally effective and still no cure either, it could happen again. In many respects it is surprising that it has not done so.

And second, sinisterly echoing the apparent invincibility of viruses, was the knowledge that no-one had yet set eyes on them. Were they like small bacteria, complex living cells on a small scale, or no more than complex molecules? No-one knew—the only clue came in the 1930s, when analytical techniques showed that the tobacco mosaic virus was an elongated structure, like a cigarette in shape, only perhaps proportionately longer.

At one stage, as we saw earlier in this chapter, it looked very much as though the use of very short wavelengths of light might prove to extend the capacity of the microscopist to observe fine detail. But the only way in which these very small particles could be seen was by the use of something entirely different. What was there? At first there was nothing; no alternative at all. But then in 1924 a physicist named de Broglie wrote a most interesting paper which was to bring about yet another change in the direction of the developing discipline. He argued that a beam of electrons (at that time regarded as essentially a stream of particles) could also be interpreted as having a definite, though very short, wavelength. It was therefore analogous to a light ray in some respects, he claimed.

Here too we may look back to find the earlier workers who reached similar conclusions; and though it is the name of Crookes which is most widely associated with the discovery of the electron beam (then known as cathode rays), it was J. J. Thomson who first realized the true nature of electrons and indeed who gave them the name.

In the following year—1899—Wiechert showed that a coil carrying a current could focus the electron beam, and in 1903 Whenelt demonstrated that an electrostatic field could do the same. Even at this early date, therefore, the notion of focussing the beam in a manner similar to that familar with light was

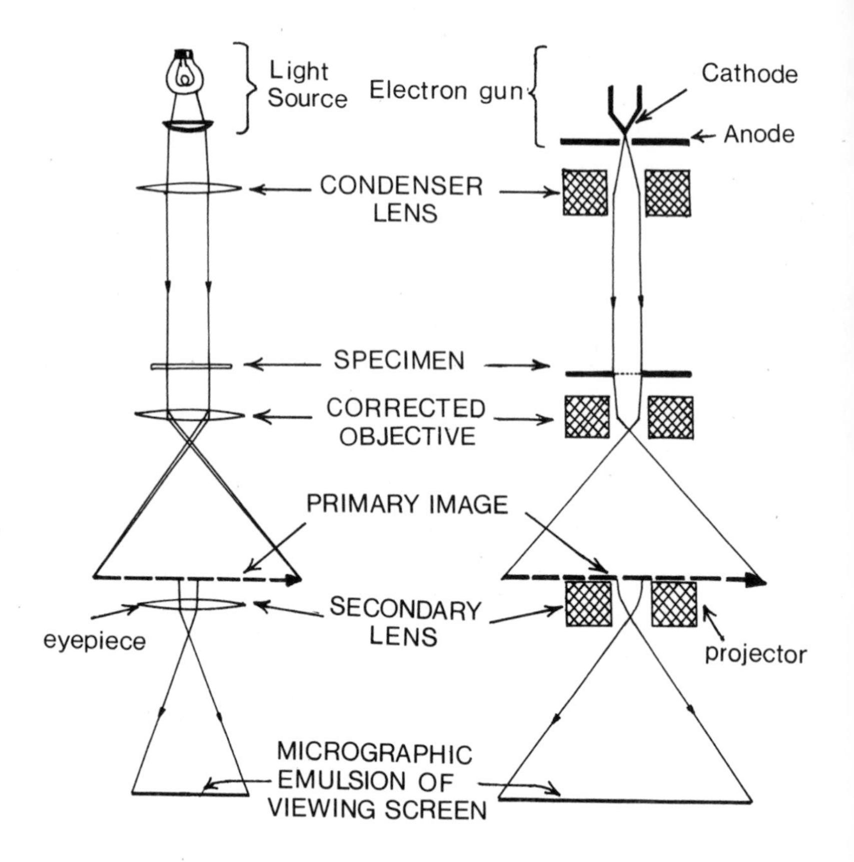

The design similarity between the optical (light) microscope *(left)* and the transmission electron microscope *(right)* is obvious from this diagram. More complex lens systems are used in practice, of course; compare with the illustration on page 156.

known. A detailed theoretical study was made by C. Stoermer in 1907 which examined mathematically the path taken by electrons passing through a coil, and even then it would have been possible to begin research leading to the still new science of electron optics.

Following on the heels of de Broglie's studies several scientists made primitive oscillographs (forerunners of today's oscilloscopes) and used them to measure electrical voltage. The beam of electrons was produced from a cathode at one end of the apparatus, and a thin yellow-green line of phosphorescence was produced on the screen at the other. The position and shape of the line changed when an external voltage was applied to two detector electrodes within.

This in itself is research in the field of pure physics, and it has little to do with the microscope at first sight. But it was the development of the oscillograph which started ripples of its own.

Working in the Berlin laboratories where one of the first oscillographs was produced were two physicists named Knoll and Ruska and in the late 1920s they began a special study of electron optics. In 1928 they succeeded in making a device which could produce a sharp, clear image of a hole in a metal plate. The size of the image was no bigger than the size of the object, and so this was not a microscope of any kind. But it did demonstrate the feasibility of focussing the electron beam to produce an image of a real object, rather than being merely brought to focus as a spot or a straight line.

In the following year they added a second lens coil to the device, and succeeded in obtaining an image that was more than the original. During the next three or four years they pressed ahead with this research and eventually could obtain images which showed a magnification of around 400 times. No-one at the time was particularly impressed, indeed few outside the field of physics would have been interested. Although the main purpose of Knoll and Ruska's work was the study of electron lenses, and had little to do with microscopy, these highly magnified images heralded an entirely new era of development for the microscope. They had invented, almost accidentally, the first electron microscope in history.

One of the main reasons why this fact took time to dawn on the scientific fraternity was that the Berlin workers were obtaining images only of the metallic outline—they were viewing shadows, in effect, of opaque objects. To study thin sections of living tissues or the structure of minerals in fine detail would necessitate the passing of the electron beam *through* this specimen.

When electrons strike a solid (i.e., an electron-opaque) material they are absorbed and x-rays are emitted. This is the principle upon which the x-ray tube is founded: here a beam of electrons is deliberately made to strike a metallic target, and the x-rays produced are then used in the normal way. This phenomenon occurring in a specimen would produce considerable damage, both due to the bombardment itself and as a result of consequential heating. Not only that, but the electron beam would perhaps produce dangerous amounts of x-irradiation which could prove hazardous to the operator. There were additional practical drawbacks; it was still not possible to ensure that the electron lenses would prove to have enough stability to enable a clear image to form, and there were formidable obstacles to the maintenance of a hard vacuum. Though the idea of high magnifications through the use of such a system must have been in the minds of many scientists in the early 1930s, there seemed to be too many practical problems in the way.

In 1933 Ruska built an electron microscope with quite a modern appearance, which has been described as the direct ancestor of today's instrument. He managed to obtain some rather inferior pictures of cottonwool fibres, but the clarity was not reliable and the specimens soon broke down in the 'illuminating' beam. But they did reveal detail that was finer than the conventional light microscope could resolve. And in the following year, Driest and Müller used the same instrument to photograph the legs and wings of a house-fly with a definition nearly five times better than the light microscope's limit. It is interesting to turn the pages of Hooke's *Micrographia* and to realize that, when the first microscopists were peering through their lenses, they chose exactly the same specimens for study.

By 1936 Metropolitan-Vickers in the United Kingdom had built the very first commercial electron microscope, known as the EM1, which might well have given rise to a prototype for mass production, or at least to a successful design for saleable instruments. But the war was too close, and over the following years interest declined. Ruska, in Germany, had substantially better luck. He went into a partnership arrangement with the Siemens company who designed and produced the first assembly-line electron microscope in the world—but the effort of the German war investment soon squeezed the sponge dry, and further backing for the project was not, for the time being, forthcoming. The Japanese, too, were beginning to investigate the commercial possibilities for electron microscopes.

In the United States another physicist, Hillier, created a successful prototype for RCA to put into production. No-one was yet sure just what the electron microscope could do, exactly. And

First experimental microscope produced by Jeol Ltd in Tokyo, Japan, in 1946. Note high-voltage generator and insulated conductors to carry the cathode potential.

nobody could claim to understand the problems of preparing specimens for examination. The problem was quite similar to that surrounding lasers, when they came into commercial production in the mid 1960s: their uses were not defined, their potentialities were largely unknown, the dangers inherent in their misuse were misunderstood.

Yet the instruments were produced, in large numbers, simply so that enthusiastic buyers could try them out for themselves and in this way find out just what might be achieved.

In just this way, there were three nationalities of electron microscope ready for widespread sale and large-scale production when war took hold on the world. The resultant pressures, both economic and political, made such research superfluous and unnecessary. It petered out.

War has a proclivity to rationalize developments in terms of pragmatic practicality. Thus it was the same stimulus that denied us, for the time being, the use of this new and exciting instrument —yet which gave mankind large amounts of antibiotics for the first time in history.

But the war effort did little to hamper the development of new optical techniques which are, after all, not so likely to consume valuable assets in the same way as the electron microscope. It was the further development of the conventional light microscope which was to bring about the next breakthrough.

At the time nothing would have seemed less likely. The science had progressed logically enough from single lenses to compound systems, from uncorrected to corrected, from achromatic to apochromatic; as has already been related we saw the absolute pinnacle of optical perfection reached and then, almost as the novelty wore off, retreated from a little. On the related front of specimen preparation, we saw how the workers in the late nineteenth century began to use coloured stains more widely, which enabled the observation of tenuous or transparent structures which would otherwise elude us. We have seen shortwave microscopy and the ultra-violet microscope; we have seen light rays used to illuminate directly or, as in dark-ground microscopy, concentrated only on the objects, which seem to hang like ghostly figures in the heavens against a velvet-black background. What else was there?

The breakthrough was called phase-contrast. Its principle is simple to grasp, though it is fair to state it is generally very badly explained. If we look at, for instance, a living cell under the microscope we will see it as a faint grey image. This is due to the fact that its structures absorb a little of the light passing through, and so the intensity—the *amplitude*—is lessened somewhat.

We are viewing a shadow. But of course very substantial structures may be of such transparency as to cast little or no shadow at all—and to the microscopist they may be extremely difficult to observe. Any transparent object, by definition, does not greatly lessen the amplitude of light passing through and in the past the only remedy for the microscopist wishing to observe such specimens was to stain them in some way or isolate them, or perhaps to use dark-ground illumination to show up the edges; none of these solutions was universally acceptable.

But, though the amplitude of the illuminating beam was not materially altered by passage through such a transparent object, there was one important change that did occur—but which could not ordinarily be detected. Light passing through an object of a different density from its environment (whether the object was transparent or not made no difference) would travel through the object at a different speed. Its *phase* would alter. So the light rays that had passed through the object would emerge sooner than, or later than, rays which had passed by undisturbed; they would be 'out of step'.

Since light is a wave form, with 'crests' and 'troughs' like any other wave, we can now see what would happen if the two beams were allowed to mingle. If they were totally out of phase, the 'troughs' of one beam would coincide with the 'crests' of the other—and they would cancel each other out. The result would be that the light in that region would be cut to a fraction of its original intensity, and the previously transparent object suddenly becomes dark. If the amount of phase difference is less than this, then the object will seem to be some shade of grey. But, most important of all, even a small difference in the optical density of an object will show up as a considerable difference in image brightness.

In practice this conferred great benefits. It meant that fine, delicately structured colourless cell organelles could be visualized in the living state, instead of being killed, fixed and stained with chemical dyes before observations could begin. Chromosomes could be seen as black ribbon-like objects moving slowly about inside a dividing cell; bacteria in the living state could cease to be tenuous flecks of almost imperceptible matter and might assume the clarity of matchsticks scattered on slate. Today phase-contrast is a vital laboratory tool for the scientist who wishes to observe life under the microscope.

Like other developments, phase contrast took time to mature. It had false starts, too.

Several workers in the late nineteenth century—Abbé, Bratuscheck and others—carried out experiments on diffraction gratings that involved phase effects, but without realizing the full implications of their work. In the early years of the twentieth century it had been discovered (among others by Rheinberg and Conrady, both of whom published in 1905) that a control of the phase of the illuminating beam could effect the contrast in the image. But it was the lot of one man, a German named Zernicke working with the Carl Zeiss company, to turn the background of theory and hesitant speculation into hard scientific fact. During the 1930s he discovered a simple and obvious way of obtaining phase control. He constructed plates of glass in which certain areas were ground thin, in the manner of a shallow trough. Light passing through these regions clearly emerges at a different phase from that passing through the rest of the *phase plate*, as it was called. Zernicke made his phase plates to retard the split beams so that they emerged one-quarter of a wavelength out of phase—i.e., exactly half-way 'out of step'. Thus by allowing the rays that had passed through the object to mingle with the direct rays the hidden structures were at once revealed. Those that retarded the light until it was exactly out of phase would seem to be nearly black, whilst those that were in phase seemed twice as bright as the background. This is only an outline of a complex technical problem, but the principle is basically simple and—as the results show—the benefits to visual observation were many.

Zernicke was another example of those specialists who came to figure in the history of the microscope from a different speciality. He was brought into the field through his work on diffraction gratings. Realizing that the phase of light could be manipulated in this way, he speculated on the effect that phase-contrast illumination might have on the appearance of an object. And one of the conclusions he reached was that slight discrepancies in an even surface could be revealed by the technique.

Dr C. R. Burch was then working at Metropolitan-Vickers on the production of accurate concave mirrors—work that eventually gave rise to the aspheric reflecting microscope, page 183—and it was by chance that a colleague of his, Van Dyke, heard something of Zernicke's work. Might not phase contrast help to check the surface of the Burch mirror, which had to be ground to within fine limits? Burch replied that he was already using a form of interference based on the same principle, though with the aim of finding a system for measuring imperfections in lenses and in the mirrors for microscopes and telescopes. No-one apart from Zernicke seems to have realized that the principle could be used to reveal the structure of tissues under the microscope. He found in his early experiments that transparent structures could be seen with great clarity, and he was quick to realize that such a device would be invaluable in the study of living material. Remember, microscopists were faced with three choices: to observe living material fresh and unaltered with a light-ground microscope; to utilize dark-ground microscopy (a method capable of revealing structures with considerable contrast, though without the subtlety of phase-contrast); or to stain with intra-vital dyes that—whilst colouring specific organelles—did not interfere with the normal appearance of the cell. With phase-contrast it was possible to examine living, transparent cells and inorganic material too, with the contrast normally associated with a fixed and stained slide.

Zernicke's debut in Britain came quietly and inauspiciously. He visited Imperial College, London, and during his stay he gave a talk to the Queckett Microscopical Club. He had with him his first phase-contrast microscope and explained how it worked. Though interest was shown, the idea showed no signs of generating enthusiasm. Zernicke did find interest shown by R. G. Canti, however, who was working at the clinical pathology laboratory at St Bartholomew's medical school. Canti was intrigued by the potentialities of the idea, but he was already overworked. Within a short while he was dead, burned out like an overloaded electric lamp. With him died the hopes of the phase-contrast microscope, at least for the time being.

Zernicke demonstrated the device to W. E. Watson-Baker of the Watson company who, as a leading microscope manufacturer, was very willing to examine any new idea that might prove marketable. But he was not impressed. He examined a test slide of the diatom *Amphipleura pellucida*, only to find that the delicate grid of markings that can be resolved by the best lenses remained indistinct. That was inevitable. Phase-contrast was not envisaged as a means of obtaining high resolution, but as a way of increasing the contrast of a specimen. Watson-Baker then looked at squamous cells from the cheek lining. This was a critical moment, for these fresh cells are seen as the faintest of images by the conventional microscope, but they stand out with striking clarity when examined with phase. He was still not impressed, and the idea was rejected.

Meanwhile, the German optical company of Carl Zeiss had begun their own development programme, based on Zernicke's earlier work. They sent one of the first prototypes to Zernicke, with compliments, as a courtesy. They did not ask him to help with the development, and they did not ask him for his advice afterwards. His unique experience was ignored. Doubtless this was a result of the political climate at that time. Zernicke had taken his discovery to England, after all, and it was up to the companies of the Reich to show that they did not need outside support. During the first years of the Second World War, Zeiss manufactured a limited number of the instruments on a production-line basis. But that was curtailed as the pressures of the Nazi war effort increased.

In England, phase contrast made one last stab at the establishment. By etching phase strips from microscope slides and accurately polishing the groove that results, Burch and Stock made a phase-contrast microscope with a high-power objective. It was ideal for studies of bacteria. The instrument was sent as a matter of urgency to Howard Florey, for use in the work on the effects of penicillin on micro-organisms. There was at that time a controversy over the action of the antibiotic, and a means of observing living bacteria affected by the drug was said to be the main stumbling-block in the path of progress towards an answer. The conditions of security that the war entailed meant that there was no news of the results for some years.

But at the end of the conflict, in 1945, Florey sent back the microscope, with his compliments. It was in its original packing. The microscope had not been used. Pressure of work, Florey explained to Burch, had prevented him from assessing it. Thus a unique aid to that vital programme of research was overlooked, and the acceptance of phase contrast received another set-back.

As the war drew to its close, Kurt Loos made a pioneer film of cell division in the grasshopper testis, which showed that phase contrast revealed the minute chromosomes as discrete objects in the living cell. Then Hughes, at the Strangeways Laboratory, made a time-lapse film of mitosis in fibroblasts. So, as long as the war lasted, phase contrast had remained unattractive, unappreciated.

But when the war ended. . . .

6 Down to the Atom

IN THE post-war generation the advances in microscopy have far exceeded all that went before. The first three centuries had slowly, almost ponderously, revealed the nature of the problem and had shown the key to its solution. But now the technical proficiency of modern science could bring its own weight to bear—and the unlocked door was thrown wide open.

The revived interest in the electron microscope was quickly rewarded; within two years of the war's end Hillier and E. G. Ramberg had manufactured an electron microscope which could give a resolution of 7Å—well over a hundred times better than that to which the light microscope could ever aspire.* Today the routine electron microscope gives a resolution of around 2–5Å, and the normal laboratory instrument uses a beam of electrons (cathode rays) from a power source rated at 100,000 volts (100 kV) or a little more.

It had been in 1932 that the resolving power of the instrument had been theoretically calculated to be 2·2Å and present-day research has shown that this certainly wasn't far out. It was in 1964 that two Japanese workers, Komoda and Otsuki, took micrographs that showed crystal lattice planes in a gold specimen—visible as fine, indistinct lines only 2·04Å apart. Lattice lines only 1·9Å have since been imaged (illustration, opposite).

Yet we must be careful about taking this too literally. There are many effects on the electron beam that high magnifications reveal, amongst them patterns caused by the diffraction of the beam by the specimen. So it is often true that the specimen itself is not being seen, only a visual artefact. In these pictures, at such

*The unit of measurement on this small scale is the angstrom unit, Å; it equals 1×10^{-8} cm.

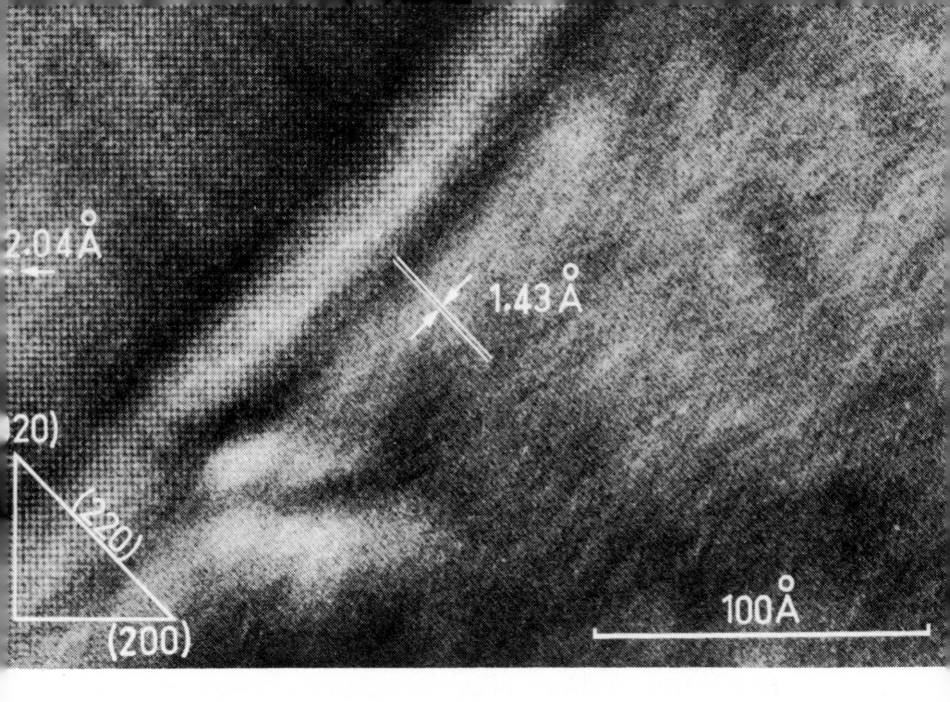

The ultimate resolution of the conventional electron microscope. Lattice planes in a gold film 0·000,000,001,43 cm apart, represent lines of individual atoms. The 020, 220 and 200 designations represent the three lattice planes. A Jeol photograph.

very high magnifications, the role of the artefact becomes important—it tends to make us 'see' more than we can in theory.

There is an interesting historical precedent for this quest for higher magnifications, of course. That is the story of the homunculus which, after Leeuwenhoek began his work, came to dominate much of the philosophical argument over microscopical interpretation. If spermatozoa were indeed the seed of mankind, then for obvious reasons there must be an entire man inside. This idea has long been ridiculed. But was it so foolish?

To begin with *there is* an entire man inside the head of a sperm, albeit in coded, genetic form. Secondly, the use of lenses at their ultimate limits and beyond inevitably introduced artefacts that no-one could foresee. There have been many more recent arguments over such cellular sub-structures as the Golgi apparatus and the ribosomes within the cell substance; arguments which bring up the same notions of interpretation as did the homunculus thesis two centuries and more ago.

And what was the homunculus? It was, as the term implies, a tiny man—a miniaturized foetus that was believed to exist inside the sperm head. Even Leeuwenhoek drew sperm that had lobed blackberry-like structures looking faintly like a much-reduced embryo, and Hartsoeker (whose design for the screw-barrel microscope we considered earlier) drew a curled-up foetus inside a sperm that well fitted the preconceptions of the theory. There is a similar precedent in the 'globulist' theories of the eighteenth century, discussed on page 95.

Our research for ribosomes amongst the granules of a cell, or our hunt for crystal lattices amongst the diffracted images on a fluorescent screen, are not so very different. And the mere fact that we now know the homunculus to be no more than a dream and the 'globules' to be nothing of the sort does not make the work at that time—*without* the benefits of hindsight—so much less worthy.

Many of the problems in using an electron microscope at the limits of its performance are purely technical—the specimen may move slightly during the exposure, and only the merest trace of shift will blur the results beyond recognition at such high magnifications, which are approaching one million times. A light switched on in an adjacent laboratory may throw the electron beam out of alignment to a slight extent, and there are always problems attached to the stability and reliability of the electrical voltages used. Add to this machining or manufacturing imperfections in the instrument's components and it can be seen how difficult it is to approach the theoretical limits of performance.

None the less we must realize that the electron microscope is more or less where its optical counterpart was a century ago—it is within striking distance of reaching its limits and we can hardly hope to go much further as matters stand.

One important aspect remains. The electron beam is able to resolve finer details because of its shorter wavelength. But the wavelength itself is a function of the accelerating voltage, so it would be reasonable to assume that a microscope working at 500 kV will have a better resolving power than a routine 100 kV instrument. There is an added benefit. Much of the damage done to a specimen by the electron beam itself is due to radiation and heating as the beam strikes the specimen. A higher accelerating voltage increases the powers of penetration of the beam, and

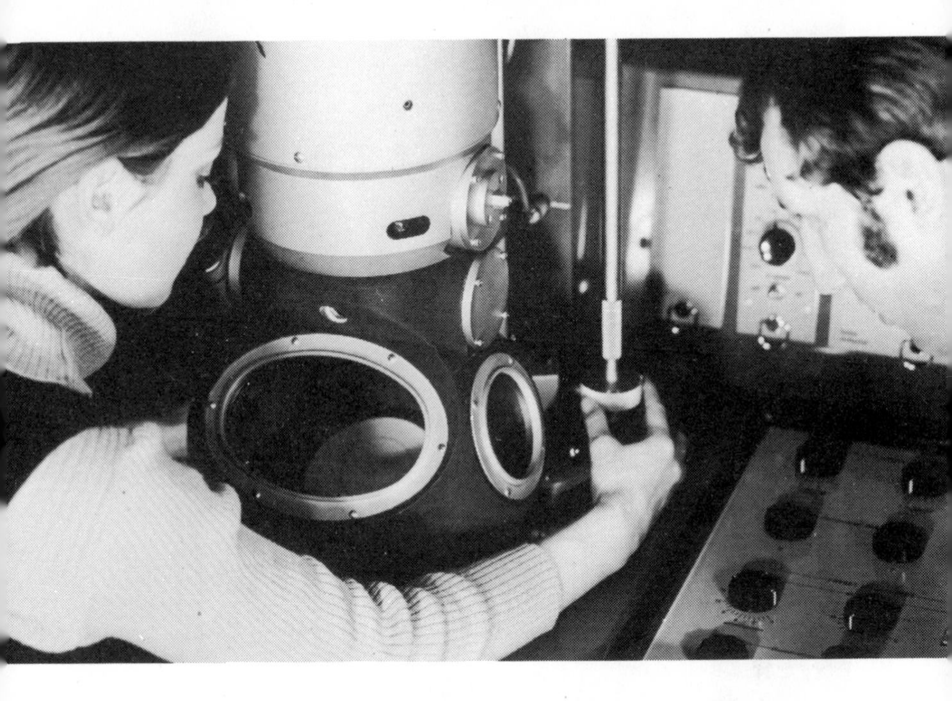

The image in an electron microscope forms on the pale fluorescent screen. The microscopist observes the screen through the glass 'port-hole' since the interior of the instrument is maintained in vacuum conditions. GEC/AEI Photograph.

would theoretically reduce specimen damage. The electrons would, in effect, shine through the specimen without being deflected and therefore without releasing such large amounts of energy as a result. This might well result in a lowering of image contrast, of course, and it would not be a panacea for all the microscopist's problems. But might it not help?

Clearly the only answer was to try and see. Several abortive attempts were made during the war years, when it was found impossible to obtain a sufficiently stable source of electricity for the cathode. In 1959 a 400 kV microscope was built in Russia and several 300 kV devices were successfully constructed in Japan at about the same time. The biggest breakthrough came in 1960, when Professor G. Dupouy in Toulouse announced the building of the first million-volt electron microscope. So powerful is the electron beam that it has been possible to examine

The vast 3.5 megavolt microscope constructed in Toulouse by the team of Professor G. Dupouy. Much of the structure contains blocks of lead to screen off harmful x-rays that would otherwise reach dangerous levels. See the illustration on page 156.

Image of an aluminium alloy containing 25 per cent silver viewed by dark-ground electron microscopy at 2 megavolts in the 3.5 megavolt microscope shown in the illustration on page 152. Only the use of very high accelerating voltages enables pictures of such specimens to be taken.

living bacteria with the instrument. They were enclosed in a thin chamber, which the beam was able to penetrate. Dr Dupouy sent me some of the first pictures that were obtained (see opposite); the intact cells, within their fluid medium, are clearly seen. Subsequent culture showed that the bacteria could still reproduce—a result that led to the confident claim that the bacteria had survived the treatment.

They may have. But it is worth emphasizing that it is certainly possible that the bacteria grown in the cultures had not been those in direct line of the beam. Perhaps they had been around the edge of the chamber, shielded from the rays by the metallic rim. However it is clear that—if the beam *did* kill the bacteria—they were certainly alive at the commencement of observations. A goal had been reached by this, the first examination of a living cell in the electron microscope. Only a few years earlier, specialists in the field had been upholding that it was 'never going to be possible' because of technical reasons: reasons which, they felt, could never be overcome. But Dupouy showed them to be wrong and the trend of opinion suddenly reversed. High-voltage microscopes became suddenly respectable with predictable results. Million-volt electron microscopes are now commercially available, and the Toulouse team have gone further to build a 3·5 MV microscope.

But the answers to all the problems do not lie in this approach. It is both difficult and dangerous to operate a microscope of this sort, since the electron beam generates x-rays on striking any metal object and these can be dangerous in the extreme.

The Toulouse microscopes are set up by remote control and are monitored by closed-circuit television, so that no-one has to be near the apparatus during the switching on and lining up operations. Only when the beam has been focussed and stray radiation has dropped to a safe limit is anyone allowed near—and even then half of the 30 ft high device is surrounded by tons of lead blocks as a radiation screen.

It is impossible to ensure an absolutely constant voltage at these very high tensions, and so the greatly increased resolution one might hope for has not been attained. But the tendency for less specimen damage, and the opportunities for examining thicker specimens than previously was possible, are themselves promising advances.

At the same time as the instrumentation was progressing rapidly there were several important parallel developments in the technical field.

Firstly there were the new methods of treating specimens for the electron microscope. In the same way that the developments in dye biochemistry had given the nineteenth-century microscopists a range of stains for their tissue sections and smears, the post-war era of vacuum physics and plastics technology threw up many new possibilities for the new race of electron microscopists. It became possible to examine solid surfaces by obtaining a replica—an ultra-thin 'cast' of the surface with all its details faithfully preserved. This was done by evaporating carbon, usually, so that it formed a thin layer over the specimen; the carbon replica became a mainstay of metallurgy.

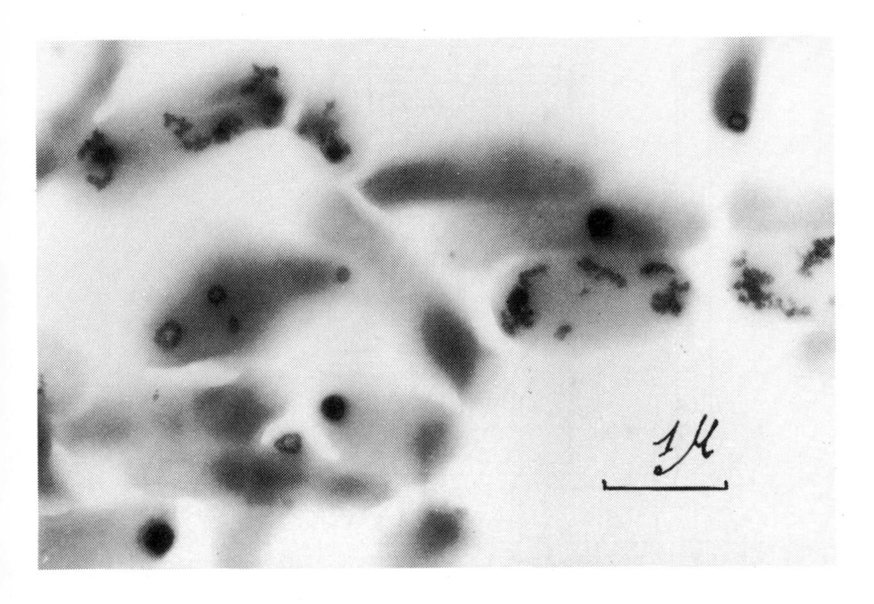

The first glimpse of life in the electron microscope. Bacteria — *Corynebacterium diphtheriae* — photographed on 7th December 1960 in Toulouse by Professor G. Dupouy. The picture, obtained with the 1 megavolt microscope running at 650 kilovolts—i.e., a little over half power—shows living organisms (or, at worst, dying organisms in a fluid environment) for the first time. All normal electron microscope specimens have to be dry.

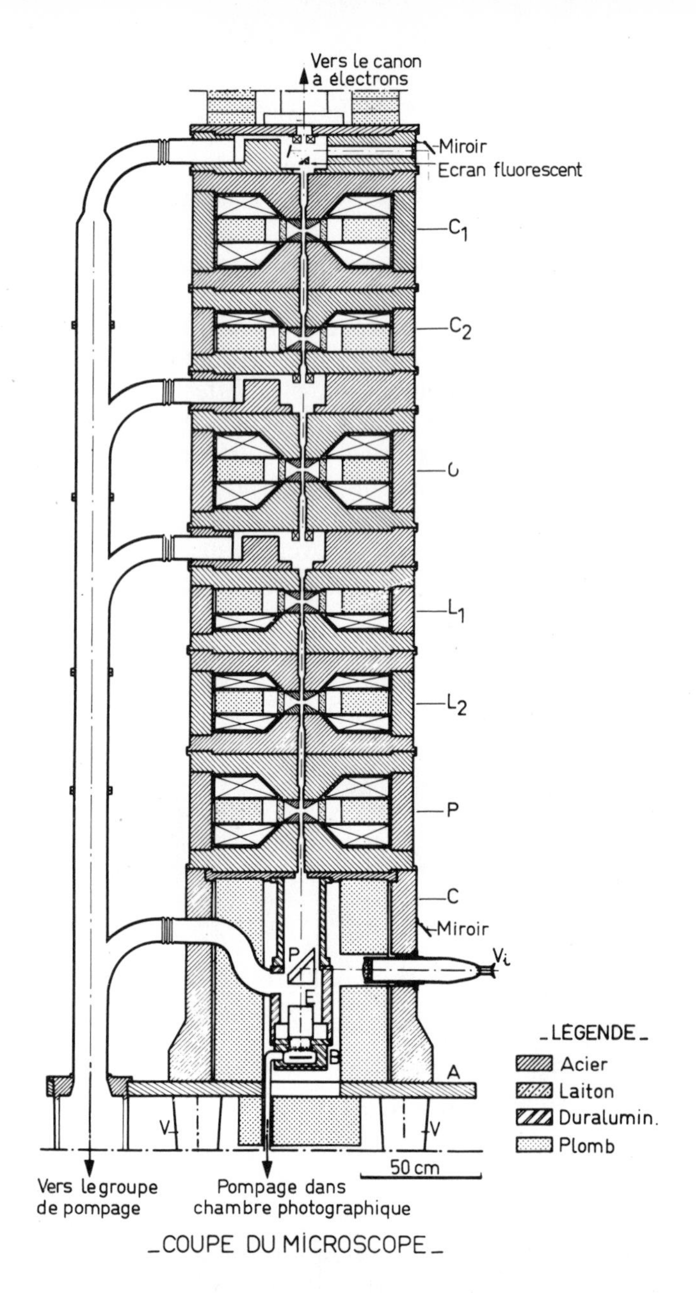

Vertical section of the Toulouse microscope shown in the illustration on page 152. Note the large amounts of steel (shaded) and lead (stippled) used to screen off harmful x-radiation released as the electron beam strikes the sides of the instrument during adjustment.

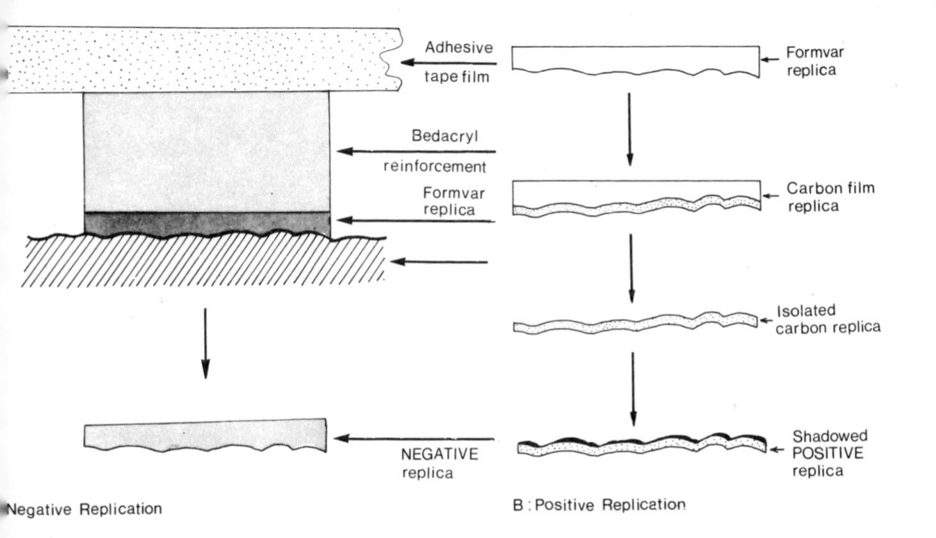

A: A popular means of studying opaque structures—e.g., a metallic surface—is the preparation of replicas. Here a plastic replica of the surface is obtained by allowing a solution of formvar to evaporate. It is lifted away by a supporting film and then treated as a 'mould' of the original surface. Higher voltages—such as those used by the Toulouse team, q.v.—enable some of these specimens to be examined as they are.
B: In cases where a positive replica is required, rather than the 'mould' shown in Fig. A, the formvar replica is coated with carbon which is then immersed in solvent (to remove the formvar) and treated as a thin film. In this diagram it has been shadowed (page 158) to reveal the surface contours in a realistic fashion. Yet we must realize that the end result—so 'realistic' to the eye—is actually a fourth-generation artefact. See the illustration on pages 162-3.

In biological studies the use of plastics came into its own. Delicate tissues could be embedded in a plastic in its liquid state. It was then polymerized and hardened into a block which could be sliced into ultra-thin sections and examined in the vacuum of the microscope without any appreciable distortion resulting.

The microbiologist found ways of evaporating metallic 'shadowing' materials so that they covered a specimen with a thin layer of metal (commonly gold or platinum). If this shadowing was done at an angle the full relief of the specimens on their background could be discerned, and a negative picture of the results threw them up from the supporting film in dramatic and vivid, almost three-dimensional clarity.

A simpler method of revealing the structure of smaller structures—such as viruses—was negative staining. It was a direct descendant of a simple technique used often in the late 1800s by freshwater microscopists. They would mix a suspension of bacteria with Indian ink and then allow the mixture to dry out on a slide. Under the microscope the entire field of view was seen to be dark brown or black—but where there were bacteria interrupting the continuity of the ink film, they shone out as clearly as stars in the evening sky. Fine flagella projecting from the bacteria would be seen as delicate, shining outlines.

The electron microscopists found that the best contrast medium to use was phospho-tungstic acid (PTA). A drop of it on a thin film covered with virus particles soon dried to leave the viruses visible in the electron beam as clear particles—even the molecule aggregates of which they were composed could be clearly seen, and the space-craft appearance of the phage viruses (which attack bacteria) was clearly revealed for the first time by the negative PTA method.

The latest new technique, freeze-etching, is capable of revealing still more new structures. By breaking open cells and tissues in this way, it is possible to reveal many unsuspected features of tissue systems. Some of the results I have seen at Berkeley, California, are endlessly intriguing, and now that world-wide interest is centred on the technique we can expect some more new disclosures in the near future.

So science has the means to handle specimens, and ways of throwing up detail with great clarity. But what exactly is the detail we are observing? As we saw previously (page 157) there is the possibility of artefact, and by the time we take a replica and gold-shadow it, then it is obvious we are looking, not at the specimen itself any longer, but an artificial outline—a remnant, a shell—of the original structure.

In observing with the electron microscope the image is initially viewed by direct projection of the 'shadow' cast by the object over the electron beam. The image itself appears on a fluorescent screen in the base of the instrument, you will recall, and where the rays strike uninterrupted they cause the screen to shine

Freeze-etched section of microvilli from the small intestine reveals 'crazy-paving' pattern of minute ducts that form part of the transport system of the cells. Photograph by Dr Bernie Gilula, Berkeley, California.

brightly in a manner essentially the same as the front of a television screen at its most brilliant. Where an opaque object blocks the passage of the rays completely, the screen remains dark. And so the observer has a graphically realistic picture of the object as a dark structure against a bright background.

It is important for us to realize that this is in principle a fortunate coincidence. The image might be recorded, for instance, as a light structure against a dark background, as a change of wavelength in the illuminating beam, as a change of polarization or any other form of alteration. Some of them (such as the last two) would not be visible to the eye, even with the aid of the fluorescent screen. The electron beam would have been modified by the object just the same, but we would need a different form of apparatus to extract the picture from the signal. It must be

Shadowed smallpox virus particles in the transmission electron microscope. This view, a *negative* print of the image obtained, is remarkably three-dimensional. Yet it is an illusion—the eye accepts the realism though the actual appearance of the specimen is similar to that in the illustration on page 168.

understood that the beam of electrons—the cathode rays (cf. illustration, page 138)—are the signal; and the object modifies it. It is this modified signal that the screen translates into something that is a recognizable picture.

That is more than mere verbal pedantry, too, for it emphasizes the essential arbitrariness of vision. We are used to seeing microscopic specimens as dark objects against a light background, and this form of presentation in the electron microscope usefully imitates our preconditioned idea of normality.

Yet we would do well to remember that the very concepts of light, dark and colour are arbitrary. All the fluorescent screen does is to make the modified signal compatible with our idea of *notional* normality.

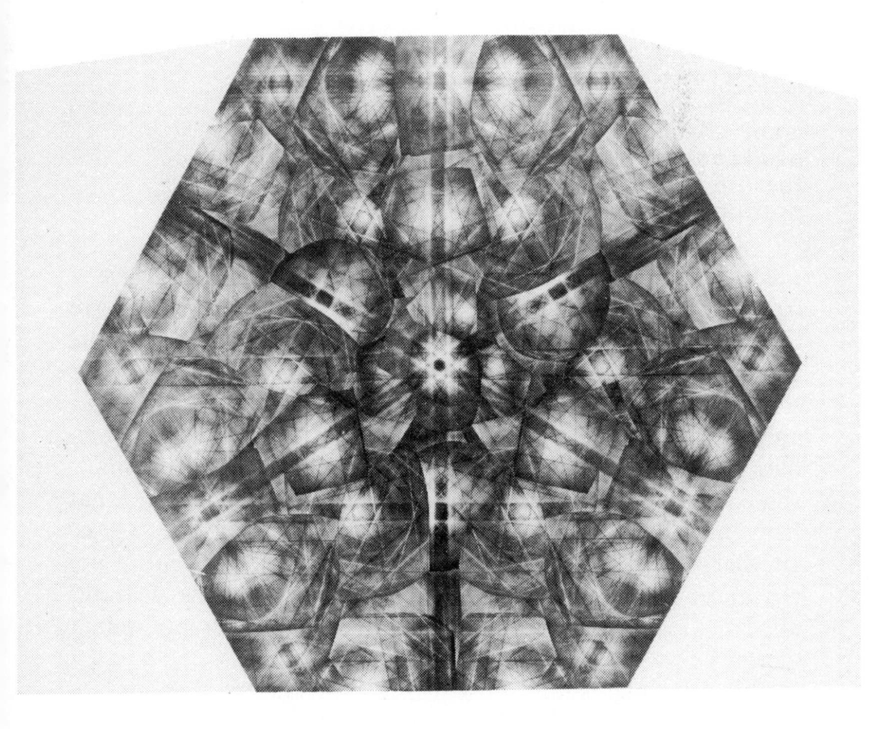

A composite Kikuchi map built up from over fifty separate pictures. These diffraction patterns, obtained by Dr Gareth Thomas's team at Berkeley, California, reveal new data on the atomic orientation in silicon crystals.

A: The true transmission electron microscope view of a shadowed sphere. No illuminant passes through the sphere, which therefore appears black; the shadow, in the lee of the particle, is clear.
B: A negative print of A. Although this is an artefact—an illusion—it at once appears like a tennis ball against a grey surface.
C: Or does it? This photograph of a ball lit by sunlight shows an important difference from B—half the sphere is *in shadow*. The eye neglects this missing feature from B in interpreting the image.

A

The human eye is conditioned into light coming from above—from the sky, reflected from a ceiling, from the sun—and we have become subconsciously conditioned into accepting this as the normal state of affairs. Shine light from below and the results are so upsetting to the brain that an emotional response can ensue, an effect used with monotonous regularity in horrific stage and film presentations. Try it, if there is any doubt in your mind: shine a torch on your face in a darkened room and observe what happens in a mirror. When the light comes from one side or another—or from above—the picture is entirely normal in appearance. But shine the torch from below, and the features appear to change and to acquire a disturbingly unfamiliar guise.

This has interesting consequences for the electron microscopist who works with shadowed particles. The eye is used to the highlight on such an object being in the upper hemisphere. Present such a picture in the opposite orientation—i.e., with the highlight at the lower pole—and the picture is now notionally

B C

upside down. It will appear as a crater, a depression, instead of as a particle. So profound is this effect that the eye can totally misinterpret the results of an electron micrograph unless it is presented the right way up—i.e., in the notionally upright orientation—when published.

Within certain limits the microscopist can use this proclivity to present his material. If he is describing certain structures as being solid and upstanding he can present the picture in the orientation which confirms this impression; similarly if they are meant to seem like functional depressions, grooves or channels, he can invert the picture and convey an entirely different impression.

The rule is simple: highlights are uppermost on 'hills'; lowermost on 'holes'.

The appearance of the shadowed micrograph itself is arbitrary too, very much more so than one tends to imagine. Consider a round, electron-opaque object which has been shadowed in the manner outlined on page 157. In the electron microscope the object itself will give a completely dark image: it is opaque, and therefore blocks the illumination beam completely. Its background of lightly shadowed support film appears a mid-grey colour, whilst the 'shadow' area where the metallic atoms did not impinge allows the beam to pass virtually uninterrupted. So we have a black image casting a white shadow—notionally the opposite of what the eye expects. Yet this is the normal appearance of a shadowed specimen in the electron microscope: something that is quite unlike anything that the eye ever sees.

Let us now consider a negative print of the same specimen; where the blacks are white and vice versa. Suddenly the picture changes. Now we see a white object casting a black shadow—a notionally normal view. At once the picture springs to life and shows us what is clearly the appearance of a white billiard ball, casting a shadow in exactly the correct way. No eye would misinterpret what it was.

Yet look closely at the picture (page 163). Does your eye reveal anything unusual? Are there discrepancies between this view and the notionally normal picture of a ball? There is one important feature of this photograph which clashes violently with the notional appearance of such an object. It is instructive —before reading on—to try to discern what it might be: but I doubt very much whether you will spot it.

The distinction between this picture and the ball is that the spherical shape of the object itself is not shaded at all. If it is a white ball, casting a shadow, then the hemisphere that is in the 'shade' should be very much darker than the highlighted part. Every solid ball you have ever observed has shown this property under all normal circumstances, yet the sphere in this micrograph does not. Instead it is an even, consistent white colour. The eye accepts it as a side-lit ball because so many of the criteria by which this kind of object is identified *are* present—the cast shadow being chief amongst them. Faced with this stimulus, the brain accepts that we are looking at a sphere on a surface; and the fact that one vital piece of the picture is missing is passed over by our interpretive mechanisms because of the weight of the other visual evidence.

This is an interesting test of our observational preconditioning, and it demonstrates the arbitrariness of visual interpretation. The 'realistic' appearance of the electron micro-

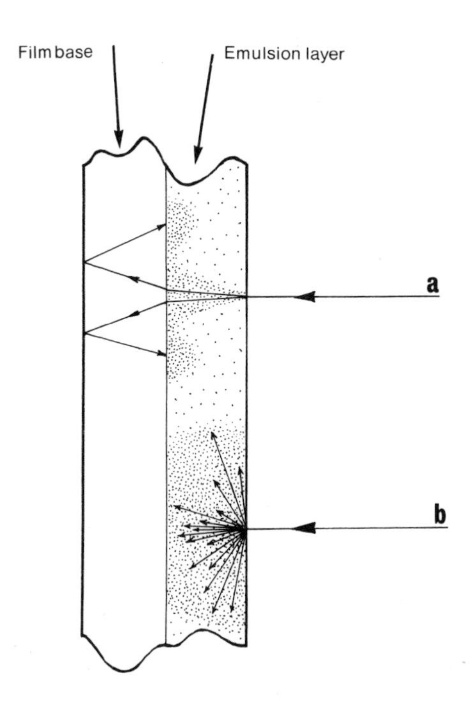

A certain degree of artefact may be introduced by the act of photographing the microscopic image. A beam of light (*a*) may be reflected from the supporting film and registered as a halo; and electrons (*b*) scattering within the emulsion may produce a degraded image.

graph can therefore be seen for what it is; a contrived simulation of certain factors that coincide with our concept of normality. The photograph has become a total artefact which the human eye sees as a mirror of reality.

The use of photographic film and plates can itself introduce further technical artefacts. The particles, as they strike the emulsion layer of the film, may be scattered; they may also form a halo. These effects, shown above, mean that even in examining a negative we have to bear in mind the secondary nature of the recorded image. It is due to the effect of an electron beam on a chemical emulsion—and only that.

So the interpretation of electron micrographs has to involve several important factors. The image itself is only a kind of chemical end-product, and its appearance is no more than an artificial impression of the altered and refined object. Our eye, and the brain that interprets its received information, extracts a great part of its inference of the nature of the specimen from the comparison between the image with our preconceived notions of normality.

And it is these, as much as the configuration of the micrograph itself, which convey to us the final impression of the specimen we began with. Often—far too often—such illustrations are published in the literature without these factors being considered adequately. The result can be confusion, or even misrepresentation, of the image.

None the less the use of these techniques has enabled us to utilize the transmission electron microscope to the best advantage. For many specimens, from metal surfaces to virus particles; from tissue sections to pollen grains; from bacteria to human cell fragments; this was the key to exploration and discovery that science had needed for so long.

But still there were important missing gaps. The main one was a desire to see whole structures—the head of an insect, say, or an entire bulky crystal—and these the conventional electron microscope could not handle.

If the specimens were too thick the beam would not be transmitted. A replica would present some of the impression of a three-dimensional object, but not the whole picture. And even if such a specimen were transparent to the beam, it would still not be seen in entirety: only one plane could be in focus at one time.

The answer was a new form of microscope, one which illuminated the specimen in the same way as the light now falling on this book as you read it. You do not see these words because of a beam of light shining through the paper, yet that is how the conventional transmission electron microscope functioned, in principle. You see these words because of a light source shining on them—daylight or an electric lamp perhaps—and your eye picks up the light reflected from the object in your hand. It shows up the threads and the patterns in your clothes too, with far greater clarity that would be obtained by holding your coat up to the sun and squinting at the outline that resulted. This was the improvement embodied in the scanning electron microscope.

Here the electron beam was directed against the specimen, and

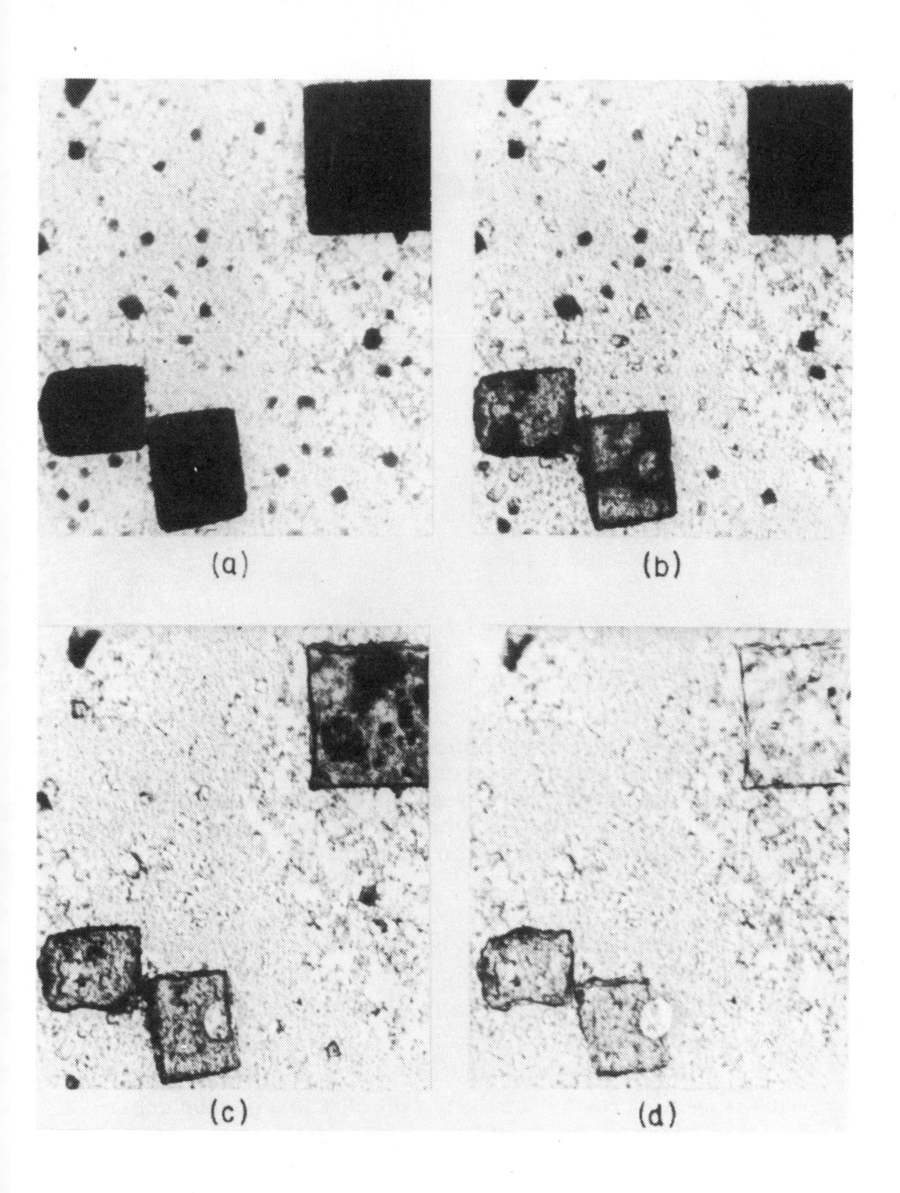

Among the first transmission electron micrographs taken in Toronto were these studies of sodium chloride crystals after different exposures to a 40 kV electron beam in the electron microscrope: *(a)* original crystals; *(b)* after $\frac{1}{2}$ minute; *(c)* after 1 minute; *(d)* after 2 minutes. Compare with illustration, page 171, which shows similar structures visualized by the scanning electron microscope.

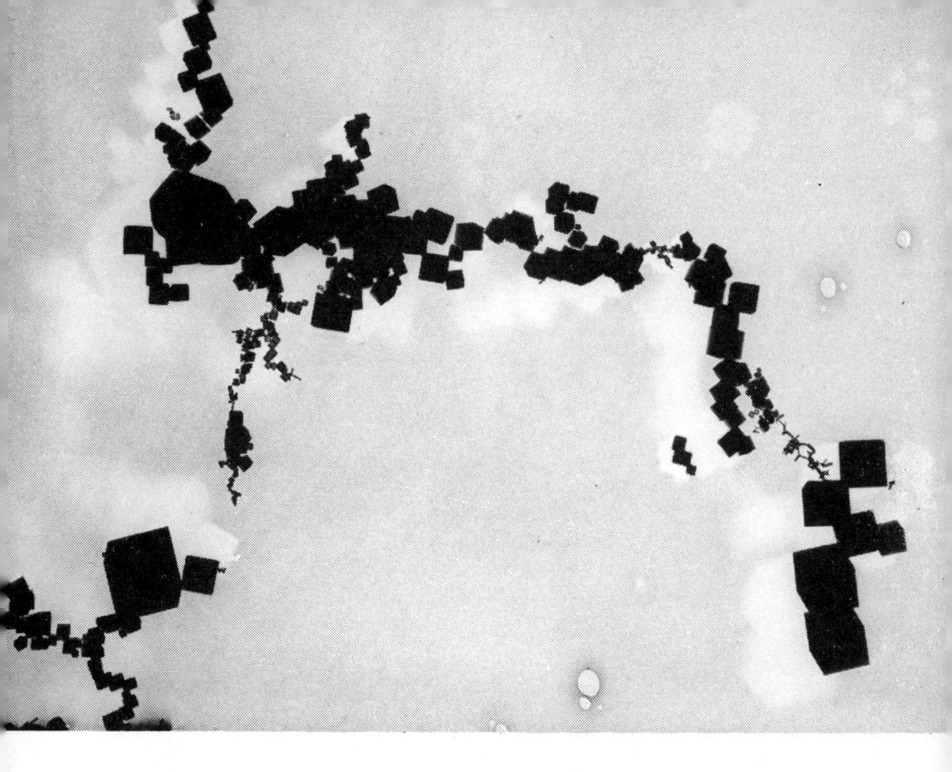

Cubic smoke particles, shadowed and observed in the transmission electron microscope. Note the *clear* 'shadows' and the *dense* 'particles'—an appearance foreign to the eye. It comes to life when presented as a negative; compare with the illustration on page 160.

scanned across it, line by line, like a kind of slow-moving television screen building up the pictures one line at a time. The photographic plate of the scanning electron microscope was exposed by the reflected electrons, just as your eye is stimulated by the reflected light; and the result was a most graphic and visually exciting form of realism. Not only this but the depth of field was enormously increased, so that whole objects could be observed and photographed.

Many of the pictures obtained with the scanning electron microscope are famous already, even to the man who has glanced at newspaper reproductions.

The view of tiny structures given to us by this newest, latest microscope is sufficiently photogenic to attract the most apathetic eye.

The great depth of field is a considerable gain over conven-

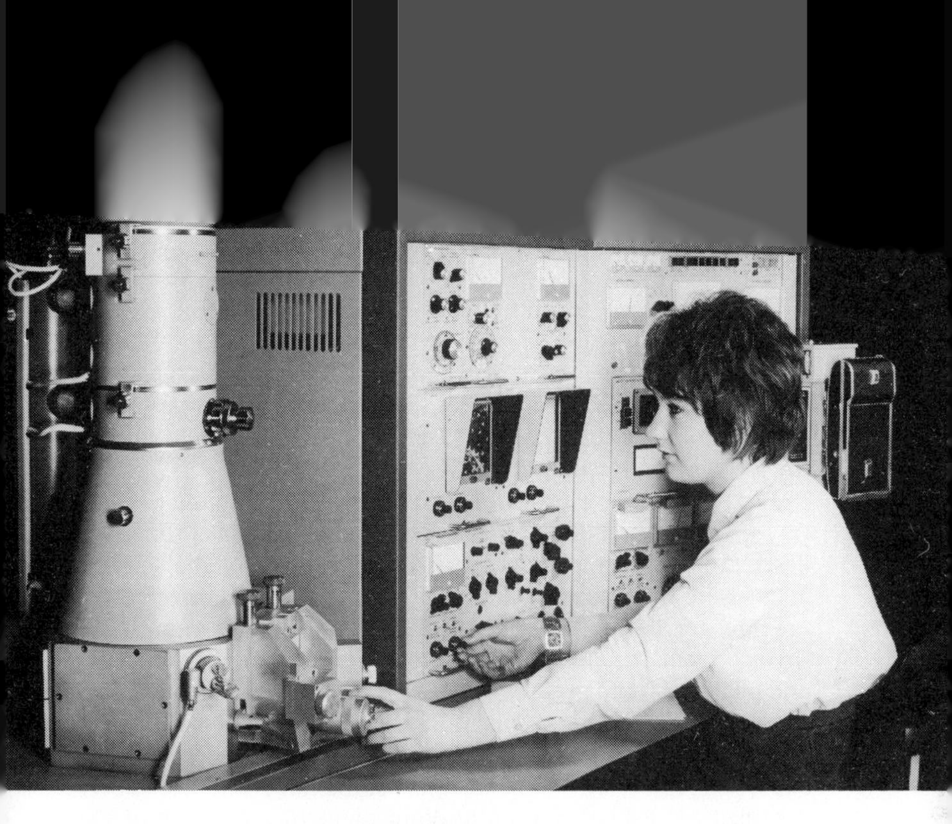

The scanning microscope in action. This photograph of the Cambridge 'Stereoscan' microscope S4 shows the microscope column *(left)*, the viewing monitors *(in front of operator)* and the camera attachment *(right)*.

tional transmission electron microscopy, though the resolving power of the scanning instrument (SEM) is far less than its orthodox counterpart. Earlier experimenters had tried a similar kind of reflected-light microscopy, when the electron beam was simply 'shone' on the specimen and the reflected beam was photographed—but this, of course, could have no marked effect on depth of field. The scanning electron beam, however, means that only a tiny part of the specimen is illuminated at one time—and so we have a kind of 'pin-hole camera' effect and virtually the whole specimen is seen in reasonable focus.

The latest refinement of this principle has been the high-speed scan—the specimen is scanned by lines, as we have already described, but at a higher rate. By the old method it took a second or two to build up the picture; now the entire picture can be scanned in $\frac{1}{25}$th second. This makes it possible to view the picture directly on a television screen and, more important, to follow events in time (such as a developing crack in a metallic foil).

The high-speed scanning electron microscope is an exciting piece of apparatus to use. A mineral surface might look like the scarred and pitted terrain of some planet; by moving the specimen slightly we seem to fly over the surface, across vast mountains and fissures that appear to be impressive, vast and mighty for all their minuteness in reality.

The turn of a single control knob zooms in the picture to some tiny detail in the middle of the field of view; it is as though we descend in a capsule to within striking distance of the surface. What were tiny granules and faintly discernible features before are now apparently towering up around us. A slow pan over the surface seems to skim lightly across the tops of the pinnacles and peaks—it is an enthralling sight. To an observer accustomed to the artificialities and uncertainty of routine electron microscopy, this is a stunningly beautiful spectacle, an unreal experience; an intrusion, almost, into something hitherto never dreamed of; a brief and enthralling flight through part of the hidden universe.

Yet for all its aesthetic merits this refinement of the electron microscope principle is scientifically somewhat valueless. We have seen new sights, but learned very little of positive value about the nature of matter and life. In a technical sense we now have the ability to demonstrate structures in a new way, and with a clarity that is impressive. But there is little to report in terms of solid advanced research of a kind that could not have been possible otherwise. In many ways the instrument is like an electric type-setting typewriter—the results are beautiful, and every self-respecting industrial complex wants to have one. But the message is much the same when we get down to it.

The scope for the future is considerable of course, and it would be wrong to suggest that it is all a waste of time. But there is a strengthening undercurrent at the moment that would have us believe that the high-speed scanning electron microscope is—like lasers, antibiotics or space-craft—a great new leap for mankind. It isn't quite that, either.

As the quest for higher and still higher magnifications has

A crystal of potassium chloride imaged in the scanning electron microscope. Note the graphic three-dimensional appearance of the solid, cubic crystal. This is a visual appearance very much more aesthetic (and easier to interpret) than the transmission electron micrograph of similar crystals shown in the illustration on page 167.

pressed relentlessly on, other microscopes have been developed that can take us progressively nearer the atom. In theory to visualize the presence of atoms in a metal specimen is not so very difficult.

Let us consider a wire specimen which has been sharpened to a very fine, tapering point. At its very end it is, let us say, a few hundred Å across. That corresponds to a few hundred atoms. The electrons in the beam are being liberated from the sites of these atoms in the cathode. Therefore, if they were allowed to impinge directly on a fluorescent screen, might not the sites of the atoms show up as bright spots on the picture?

That is how it would seem in theory. But the practical limitations are a handicap. The electrons do not move in precisely calculable trajectories, but tend to wander; and the result is an out-of-focus pattern with a maximum resolution of around 10Å—not as good as that of the conventional electron microscope. So the field-emission microscope, as it is called, has had only a limited career. But its inventor, Edwin W. Müller, took the idea a stage further. He realized that the electrons liberated from the source are too easily scattered to make a properly focussed image. So he decided to utilize a more massive particle instead—one which would be less easily deflected. The result of investigations along these lines have given us the field-ion emission microscope. Its functioning is simple to grasp.

The object wire (carefully constructed to be only a few hundred atoms across) is given a strong positive charge of electricity.

Meanwhile a very small concentration of helium gas is admitted to the vacuum of the microscope itself.

The atoms near the charged emitter, as the source wire can be known, each lose a negatively-charged electron which is attracted towards the emitter. Thus each helium atom is changed into a positively-charged helium ion. The positive charge on the emitter strongly repels each of these ions.

Each ion of helium is forced away from the atom in the emitter which repels it. Since the helium ion is more than seven thousand times as heavy as an electron, it has less of a tendency to wander from its straight-line course and therefore the spot at which it impinges on the screen has a direct relationship to the position of the atom which repelled it from the emitter. The crystalline patterns taken up by the atoms are clearly seen and

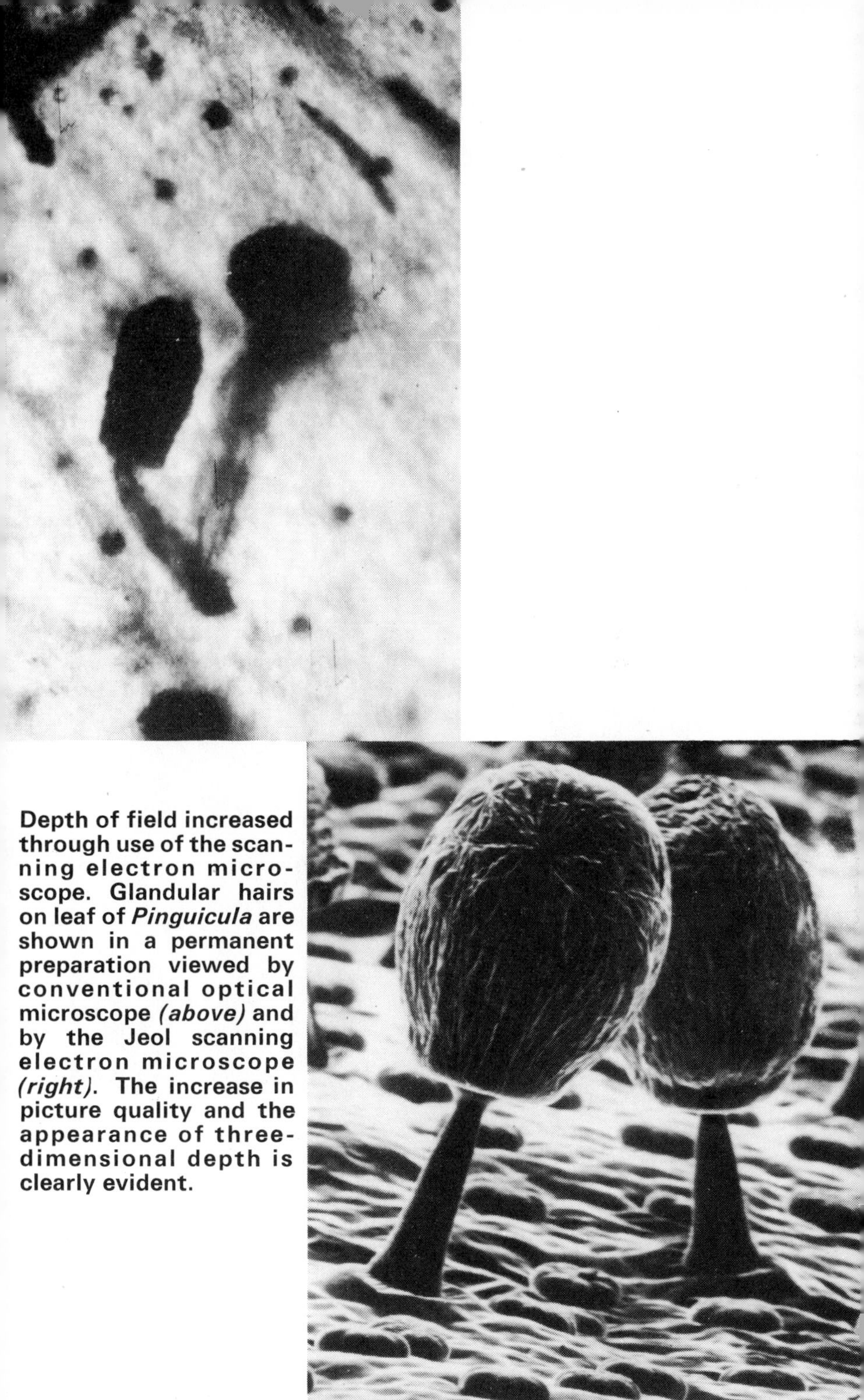

Depth of field increased through use of the scanning electron microscope. Glandular hairs on leaf of *Pinguicula* are shown in a permanent preparation viewed by conventional optical microscope *(above)* and by the Jeol scanning electron microscope *(right)*. The increase in picture quality and the appearance of three-dimensional depth is clearly evident.

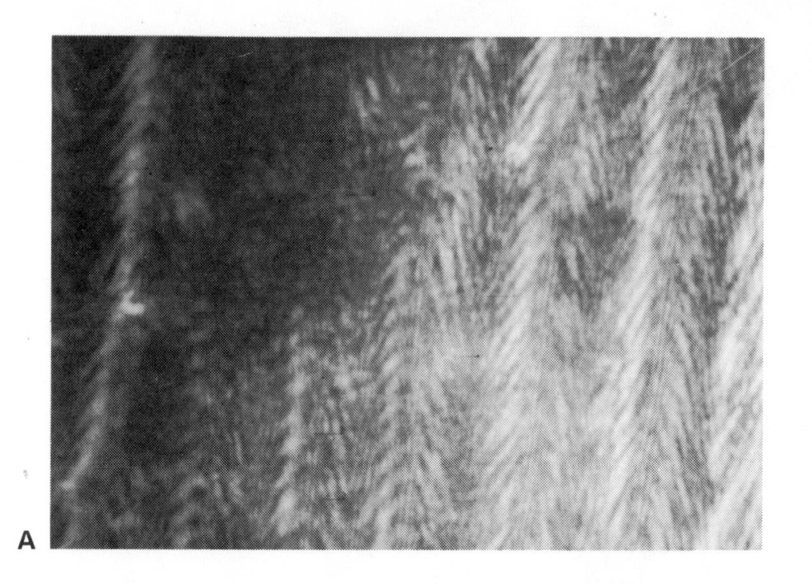
A

B

The improvement in microscope image over three centuries. Studies by the author.
A: View of a feather seen by simple bead lens of the type used by many pioneer microscopists.
B: Higher magnification and greater clarity were achieved by compound microscopes in the early nineteenth century.

C

D

C: This high-power view of the central region of B gives an indication of the best performance obtained from a modern optical research microscope.
D: Study of part of the central region of C by the Jeol scanning electron microscope shows the great clarity, depth of field and realistic appearance that the method provides.

the small bright dots are believed to correspond to the positions of the atoms themselves.

The most recent development in this field has been some startling research at the University of Chicago. By combining some of the features of the scanning electron microscope and the field-ion emission instrument, it has been possible to take photographs at magnifications of up to five million times. This has revealed chains of thorium atoms, and atoms of uranium too, as distinct images.

The instrument has been dubbed a 'scanning and transmission electron microscope'—STEM—and is already due in commercial production. The new STEM was developed by a research team headed by British-born Professor Albert Crewe in the physics department of the University of Chicago. An ultra-fine tungsten cold cathode field emitter provides the source of electrons which are attracted by the focussing effect of a large electric coil into a beam only 5 Å across—the size of a few atoms. One of the first experiments carried out by the Chicago team was the resolution of uranium atoms. They were seen as small, indistinct particles on the photographs, and there was considerable controversy over the interpretation of the results at first. Subsequent analysis and confirmation by other workers has confirmed the nature of the photographic image.

Crewe's work means that we have the capacity to develop an entirely new branch of microscopy, using the benefits of the high-resolution TEM and the greater depth of field of the SEM. When I recently saw the microscope in action at Chicago, perhaps the most impressive aspect of the device was its link-up to equipment in the laboratory next door. A range of analytical facilities can be linked to the STEM. The contrast can be controlled electronically, specific measurements and analyses carried out, counts, digital processing by computer and so on . . . the information that can be extracted from the magnified specimen is considerable. In the few years that have elapsed since the microscope was first announced, a dozen or more similar projects have begun at laboratories across the world. Manufacturers have shown interest, too. A second version of Crewe's original microscope was made by AEI at Harlow, Essex, but production-line manufacture has been postponed.

Development costs of such a revolutionary instrument are

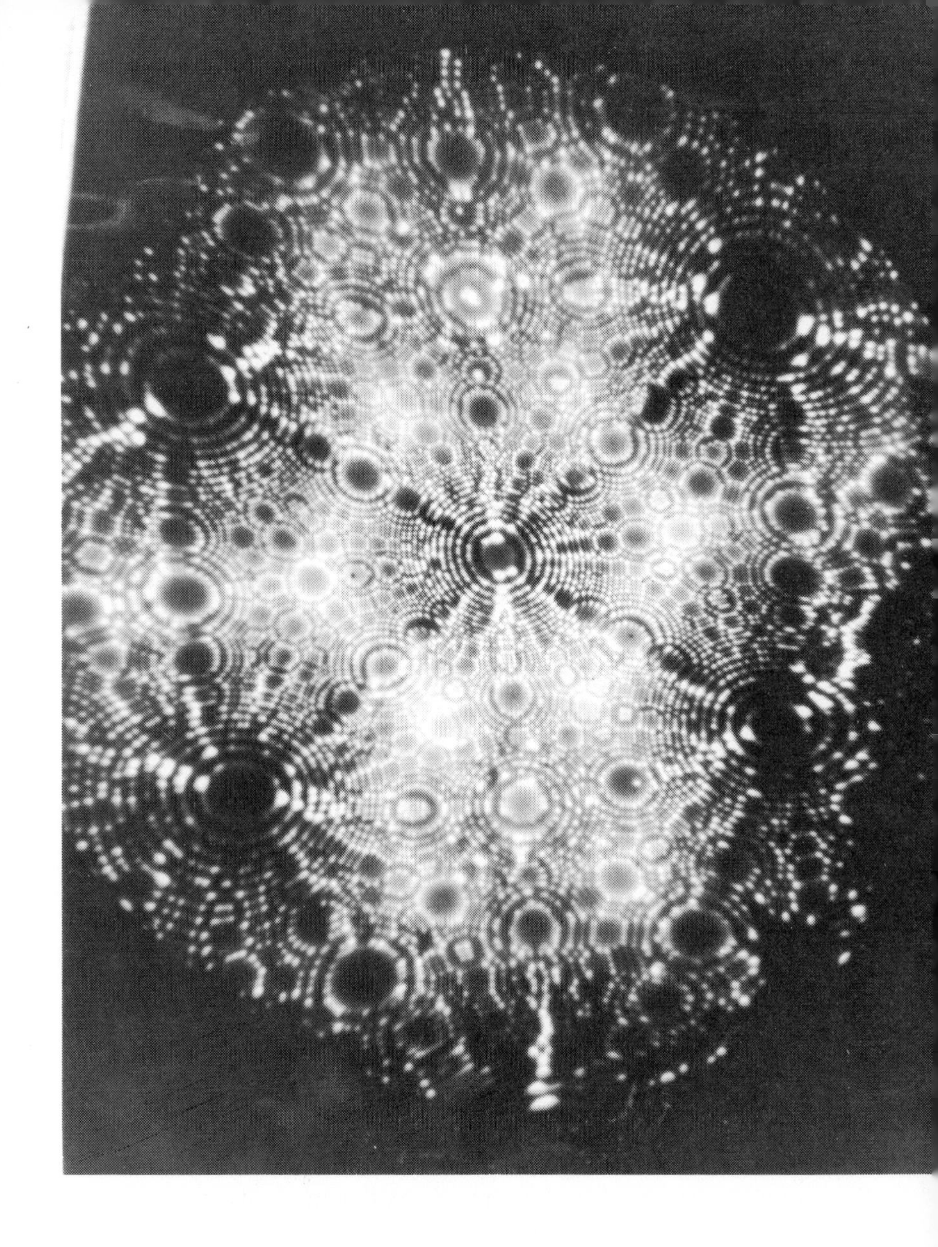

Atoms of tungsten revealed by the field-ion microscope. The atomic pattern at the tip of a wire specimen only 350Å in diameter imaged with helium ions at the temperature of liquid neon. This picture, by K. M. Bowkett and D. A. Smith of the University of Cambridge, appears in their book *Field-ion Microscopy*, published by North-Holland.

very high, and at present it is doubtful whether there would be enough demand to make the step worth while. The company that developed the first scanning electron microscope, Cambridge Scientific Instruments Ltd, is considering the matter as this book goes to press.

There can be no doubt that the STEM is a profound development. In some respects it is the ultimate refinement of the electron microscope principle, and its role in helping us to understand the hidden universe is potentially enormous. In ten years' time we may see Crewe as a Nobel prizewinner, and his design as a fundamental item of every laboratory's equipment.

At present—though no-one doubts the uses and value of the STEM—it is remaining largely ignored solely because it is not yet fashionistic. It is the story of the phase-contrast microscope, electron optics, the cell theory, *seminaria* and fermentation over again. When we can stop to consider the criteria by which we recognize scientific merit, and the frustration that inevitably follows the waiting-for-Godot mentality that has dogged our heels for aeons, it may be that exciting, important, relevant developments can incorporate themselves into the framework of civilization more rationally, and with greater value. For the moment we must—once again—wait for the bandwagon to start.

So in three centuries we have moved from the first glimpses of bacteria to the direct visualization of atoms themselves. In terms of magnification there is little more for which to hope. But there are still refinements awaiting discovery. And there is one further type of illuminant we have yet to grant a fair trial.

The electron beam is not the only sort of radiation with a wavelength sufficiently shorter than visible light to enable high resolution to be attained. X-rays can also be employed, and they travel very much further through air than electrons, making large vacuum assemblies unnecessary. The device is almost incredibly simple. It utilizes a very small point-source of x-rays and the object is merely placed between the source and the screen. The smaller the distance of the object from the source, and the further the screen, the greater the eventual magnification.

The limiting factor is the size of the x-ray source; and to date the resolution capacity of the x-ray microscope is not appreciably better than that of the light microscope. But it can be used to examine structures through which light will not easily pass;

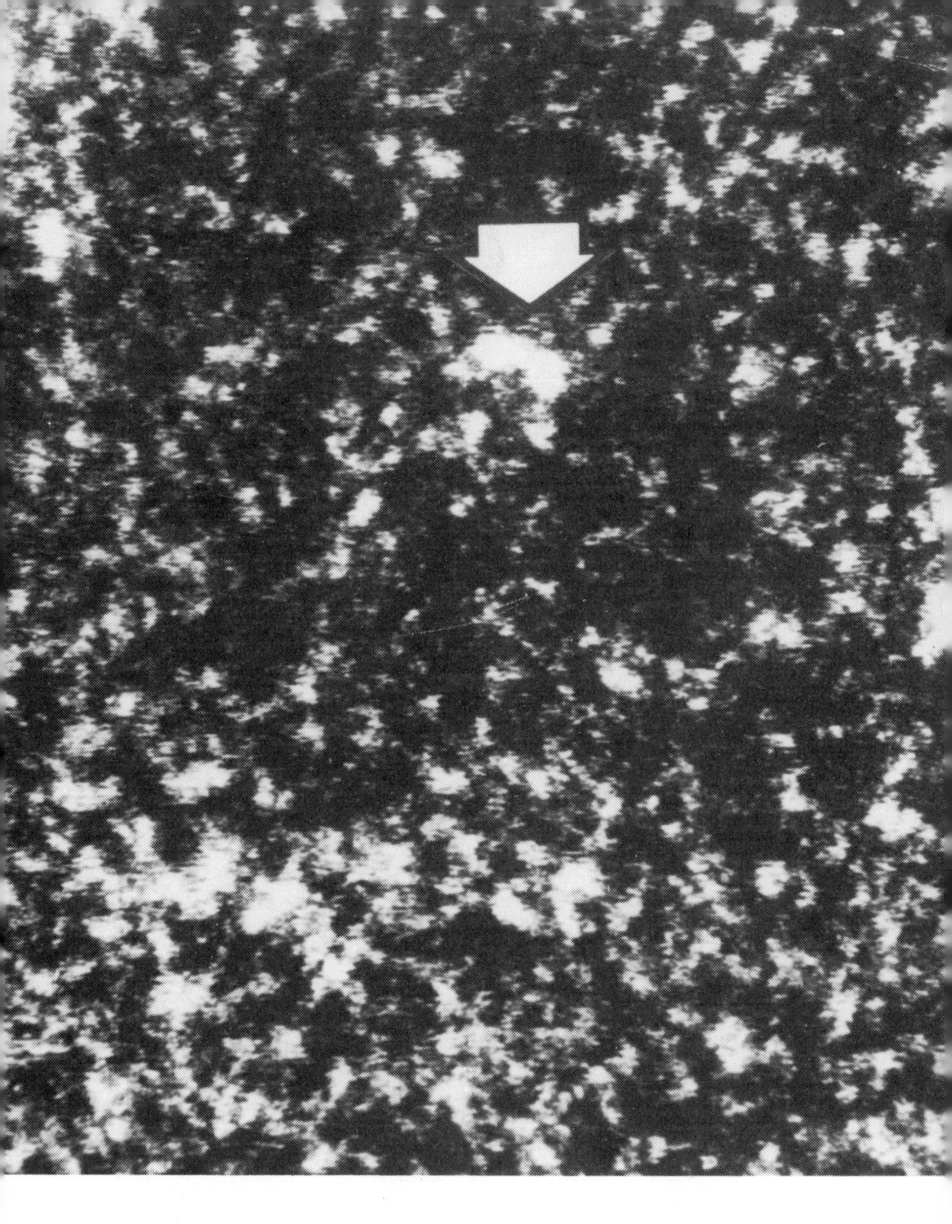

Controversy surrounded the first micrograph showing single carbon atoms (the irregular masses arrowed) obtained by Albert Crewe at the University of Chicago. His designs for electronic microscope systems are very advanced, and the interpretation of the image is often complicated. The evidence that Dr Crewe has, strongly suggests that these are indeed the 'shadows' of individual atoms.

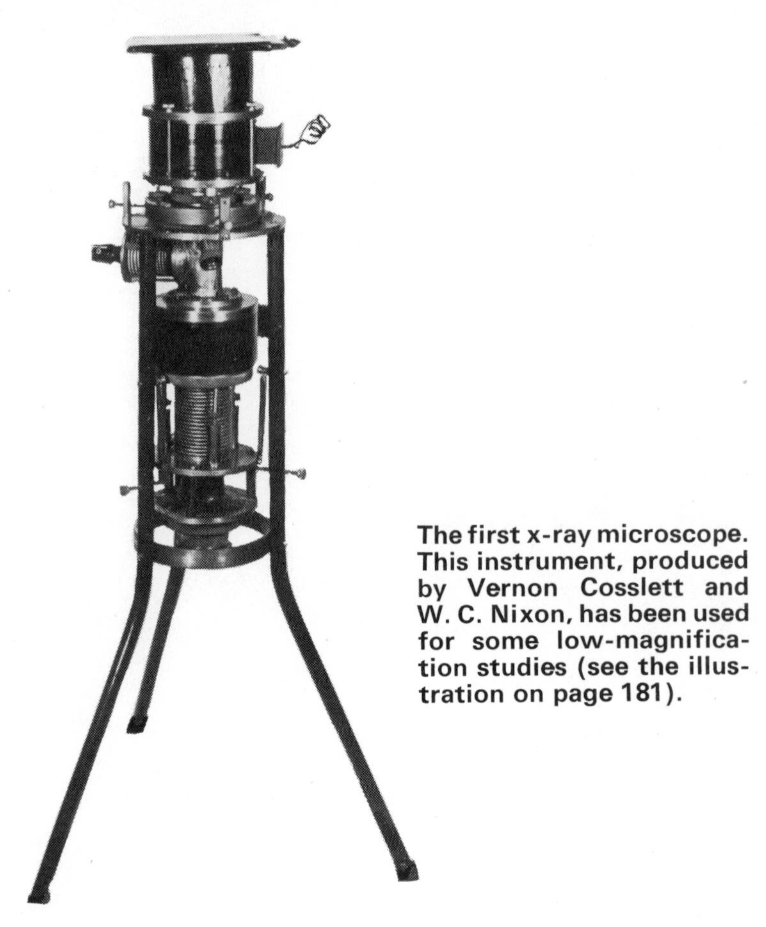

The first x-ray microscope. This instrument, produced by Vernon Cosslett and W. C. Nixon, has been used for some low-magnification studies (see the illustration on page 181).

blood circulation in the ear, the interior structures of insects, etc. It is still a little-known form of microscope. It has been said that the x-ray microscope could not be focussed by lenses, but it is possible that the rays, which ordinarily penetrate glass with ease, could be made to 'graze' off mirrors and perhaps focussed as a result. Perhaps this may give us a practical high-magnification x-ray microscope. But this is a specialist interest, and is only a very small part of the modern field of microscopy as a whole.

Between the rays of visible light and x-rays lies the zone we define as the ultra-violet wavelengths. As we have seen on page 122 there are possibilities attached to the use of ultra-violet as an illumination source for microscopy and post-war research has realized many of them. They are, however, of only limited

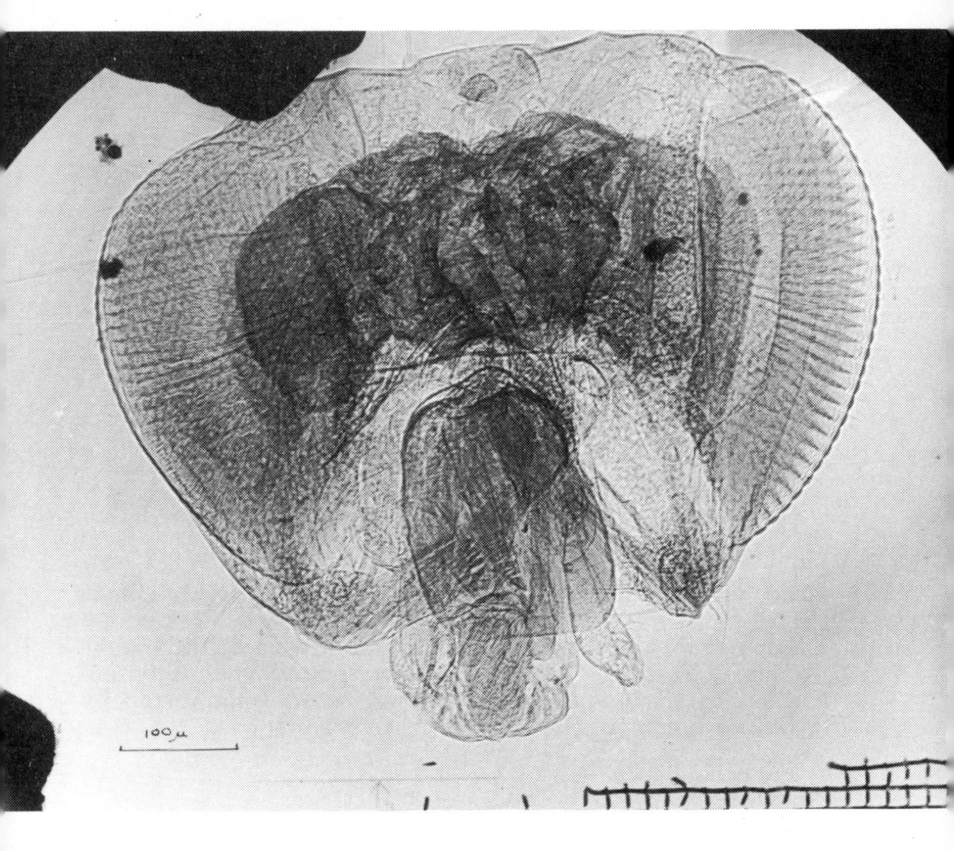

Head of *Drosophila*, the fruit fly, imaged in the x-ray projection microscope by Cosslett and Nixon at the Cavendish Laboratory. The internal structure of the eyes is revealed in a unique manner.

interest, as the very best results are only about twice as good as those of the light microscope and the capacity of the electron microscope to resolve very much finer detail is far greater than this.

There is a further type of optical microscope which we should mention. It is a form in which ultra-violet rays could be focussed as easily as light, in which there could be no chromatic aberration and in which a change of illuminant from, say, red light to ultra-violet would require no change in focus. Furthermore, instead of the objective lens crowding close to the specimen, there can be a considerable working distance. That, when hot

The author adjusting the objective mirror of his Burch reflecting microscope—the largest optical microscope ever put into production. The twin controls of the mechanical stage can clearly be seen below the objective housing; the heavy condenser mirrors are supported by the triangular bracket just beneath the stage.

specimens are being observed or when micromanipulation is to be carried out, would be a significant advantage. Yet in spite of these factors this form of microscope is only rarely heard of.

It overcomes the drawbacks inherent in the conventional microscope by utilizing mirrors instead of lenses.

This is the principle of the high-power telescope, of course. The largest astronomical telescopes in the world utilize mirrors as reflecting elements instead of lenses. But here the problems are different. Such a long-focus mirror is mostly flat, and it is shaped into a gentle saucer-like profile by grinding. It is also very large, so that a very small discrepancy would not matter unduly. But lenses for microscopes have a short focal length because of their very high power, and mirrors would have to be very carefully produced to match this level of performance. The technical problems are considerable. The earliest attempts at designing a reflecting microscope date from the time of Newton (page 93) and around 1837 the Italian microscopist Amici demonstrated a quite successful model.

There have been several more recent designs for reflecting microscopes. A high-performance instrument announced in 1947 was C. R. Burch's precision-ground microscope shown on page 182.

Shortly afterwards a reflecting objective, designed by C. G. Wynne, was produced. Cleverly it was designed so that the reflecting surfaces were separated by glass. A simpler project was constructed by J. Dyson in 1949, and allowed the working distance between a conventional microscope and its object to be greatly increased. The prime application for this device was the examination of hot metal surfaces which might otherwise have damaged the microscope itself.

Many of the forms of reflecting optical system have been made with spherical surfaces, for ease of manufacture. But for accurate magnification without spherical aberration the surface should have a more complex profile. In the 1950s, as part of a commercial programme which resulted in ten large reflecting microscopes being produced at Bristol, C. R. Burch designed a complex aspherizing machine to grind the mirrors to a profile correct to less than an optical wavelength. Yet the practical problems remain, and for all its potentialities the reflecting microscope has never really arrived.

More sophisticated forms of optical microscope utilizing the phase principle have, however, come to dominate the scene. The phase contrast instrument has been developed to a high state of perfection.

The drawback to conventional phase contrast is the fringe of interference that surrounds any large object, and the obscuration that results from overlying structures. However a recently developed sophistication allows a very narrow plane of focus to be observed—and the benefits for certain applications are clear. A sophistication of the technique—known as interference contrast—enables accurate qualitative measurements to be taken of the structures themselves.

In Poland, a Warsaw scientist, Dr M. Pluta, has recently developed a form of what he calls amplitude contrast. In this technique the amplitude of the two beam systems is altered, and not their phase; this is carried out by the use of specially hardened soot rings placed more or less where the annuli are situated in the conventional phase contrast microscope. He

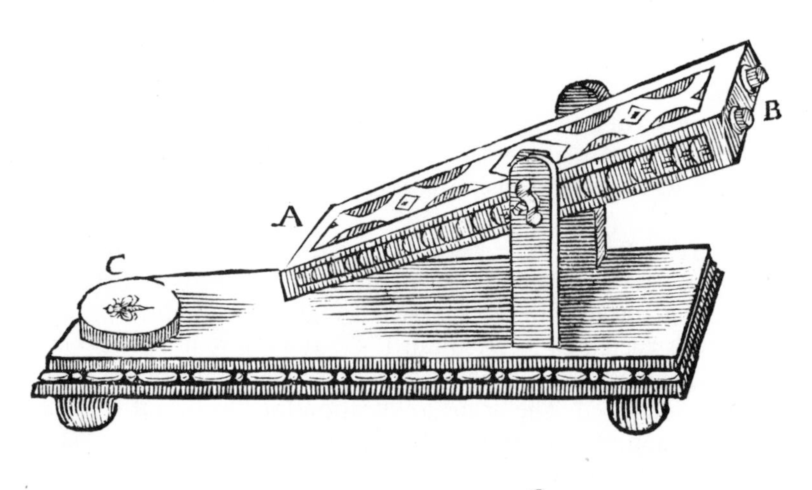

The Chérubin d'Orléans microscope showing the binocular principle. It dates from about 1680.

claims that the contrast for objects that are weakly stained (i.e., that have some amplitude change already) can be greatly increased. The sharpness and quality of his results shows a gratifying increase over phase contrast, and this new method could become more widespread in the future. Another approach, Nomarski contrast, gives a dramatically three-dimensional appearance, relying on the notion of preconceived normality we described on page 162. The results are pure artefact, however.

Even the conventional optical microscope has been developed to give a whole range of refinements. The greatest drawback—lack of depth of field—can be overcome by the use of a very sharp slit of light exactly in the plane of focus. The specimen can be moved relative to the microscope so that only the area that is being illuminated is recorded on the photographic plate—and that region is, for obvious reasons, bound to be in focus. In this way a very irregular specimen can be successfully photographed with great clarity and sharpness.

Ciné records of cut sections also throw an interesting light on the interpretation of solid structures. In this technique a series of thin sections is cut from, say, a plant stem and the mounted specimens are then photographed on ciné film.

The projected image gives the impression of travelling down

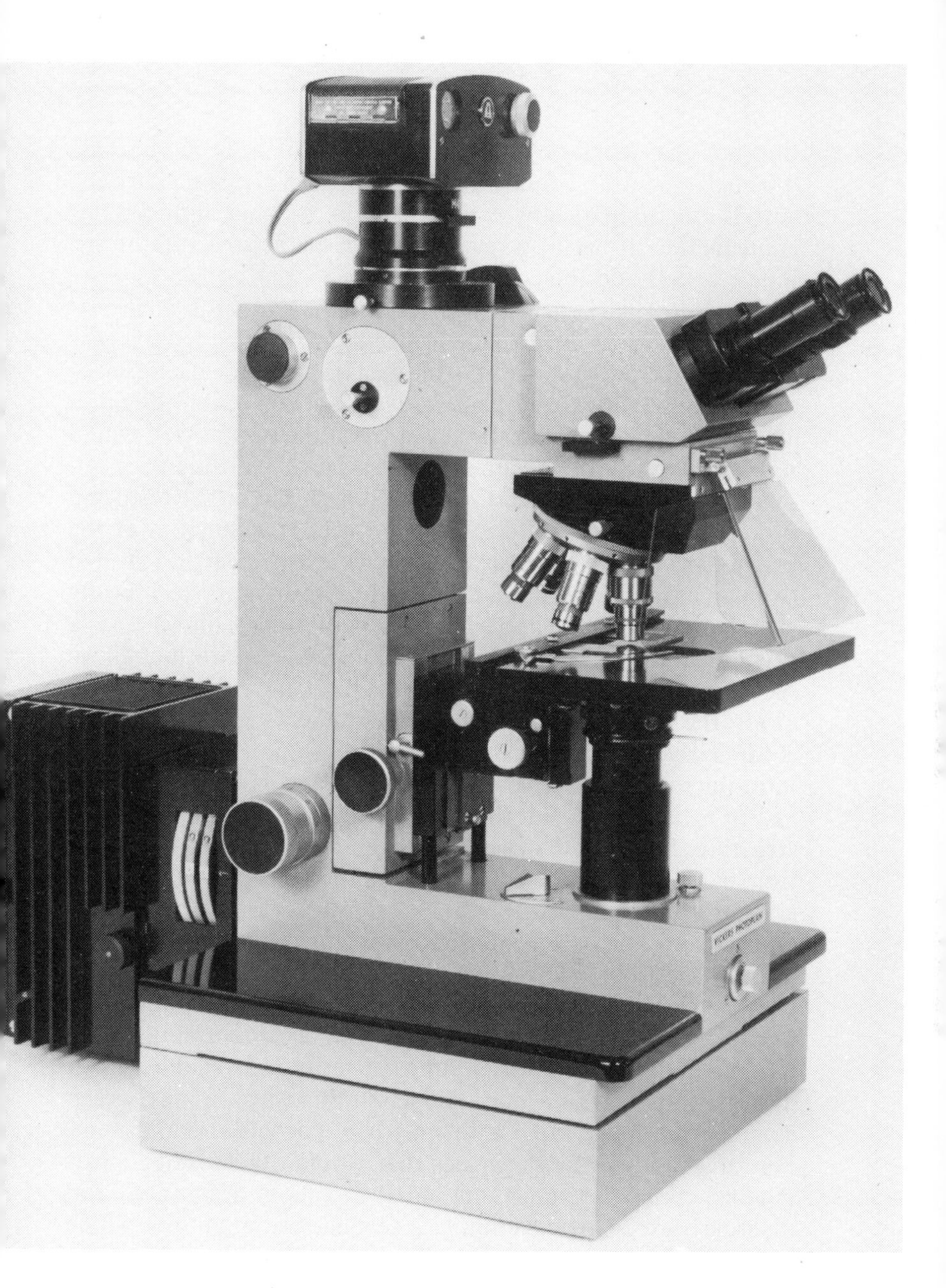

Modern research microscope—the 'Photoplan' by Vickers. Note focussing stage; low-level controls; box-like limb facing *away* from microscopist; and plastic shield for use in ultra-violet fluorescence microscopy.

through the stem, observing the shapes of cells as we pass 'through' them and following the course of the fluid-carrying vessels in a dramatically realistic fashion. As a teaching aid this is invaluable. It has also revealed much of the structure of fine blood-vessels and the courses they take.

The use of dark-ground microscopy (page 112) is now relatively rare. This is unfortunate, since light reflected from a very small structure can make it visible against a dark background (as we can see tiny dust or smoke particles in a beam of light in an otherwise darkened room) even when the object itself would not be resolved in the ordinary way. The size of the image observed depends *not* on the size of the object, but on the resolving power of the lens—the size of the object is a function of the brightness of the image instead. Thus we are seeing light reflected from an ultra-microscopic structure, but we are not—technically—seeing the structure itself. If it is a small particle all we see is a tiny dot: the object may be round, elongated, square; but all we see is a small disc of light. Obviously this is not truly an 'image' at all; the technical term is an 'antipoint'.

Similarly we cannot truly speak of 'resolution' at this level of ultra-microscopy. The term I have proposed is 'to visualize'—i.e., to make visible. But this technique, since it can reveal some structures beyond the normal 'small end' of the limits of resolution, can provide the most detailed information about the presence of very small structures.

And finally we have witnessed a mushrooming proliferation of ancillary aids to microscopy. The attachment of a television camera to microscopes a score or more years ago was seen as an aid to teaching and demonstration. But the processing of the electronically coded information that results has enabled us to attach computers to a microscope. In this way we may screen samples and make rapid assessments of particle size and number, areas of specific structures that would otherwise be incalculable, count the number of small objects in a confused and disorderly array in a fraction of a second, and monitor a whole range of microscopic phenomena with accuracy and repeatability. The feeding of this kind of information through computer programmes will, no doubt, become increasingly widespread.

Thus the microscope has finally arrived.

And, as this survey of the available apparatus reveals a strikingly varied armoury of a very sophisticated kind, so too the applications in practice are very varied.

For most people, indeed, the microscope symbolizes science: it is the by-word for research at a high level, and it is a word that crops up whenever scientific research is brought to the public attention. This is not so far divorced from reality, when we go into it, for the ramifications of microscopy throughout the modern world are widespread and profound. Microscopes are amongst the vital equipment of the doctor, the art historian, space scientists, the detective, and workers in a host of technological fields. The breadth of scope is greater than most people imagine.

For instance in industrial use the microscope has had a profound effect on traditional technologies. The triode valve of the post-war years—as big as a light-bulb—is now replaced by tiny units the size of sand grains. The microscope has made this possible. Whole amplifiers can now be produced that are smaller than a matchstick, and complex electronic units are now built up on tiny slivers that have to be fitted into a metallic container for ease of handling. Even though the finished component is still diminutive it is worth remembering that the bulk of the unit is merely its housing. The electronic hardware is no more than a delicate square of plastic in the middle—and even that may have been tested in a microscope designed for the task.

Miniaturization is more than technical showmanship. Without these minuscule components we would have had to forgo much of our modern radio, television and radar equipment. A range of medical devices from cardiac pacemakers to remote-control detector 'pills' would have been rendered impossible, and others—such as portable ECG machines used in emergency cardiac cases—would be impractical. We would not have had those tantalizing glimpses of the Moon and Mars, and men would yet have to wait for that first walk on the lunar surface.

The scanning electron microscope has brought an added benefit to the electronics industry. Because the image that we see is, in essence, an electronic signal it can be used to reveal information that is not otherwise easily obtained. For instance, the application of a voltage to a small component will show up readily as a change in the 'visual' appearance of the unit under the microscope in the factory. And so detailed, minute assessments of performance can be undertaken and for the very first time the technician can literally see the electrical charges at work. He can visualize a voltage right there, in the component.

Everyone knows of the use of microscopical technology in the metallurgical field. The space-age alloys are studied in this way as an important part of predicting their properties. The new

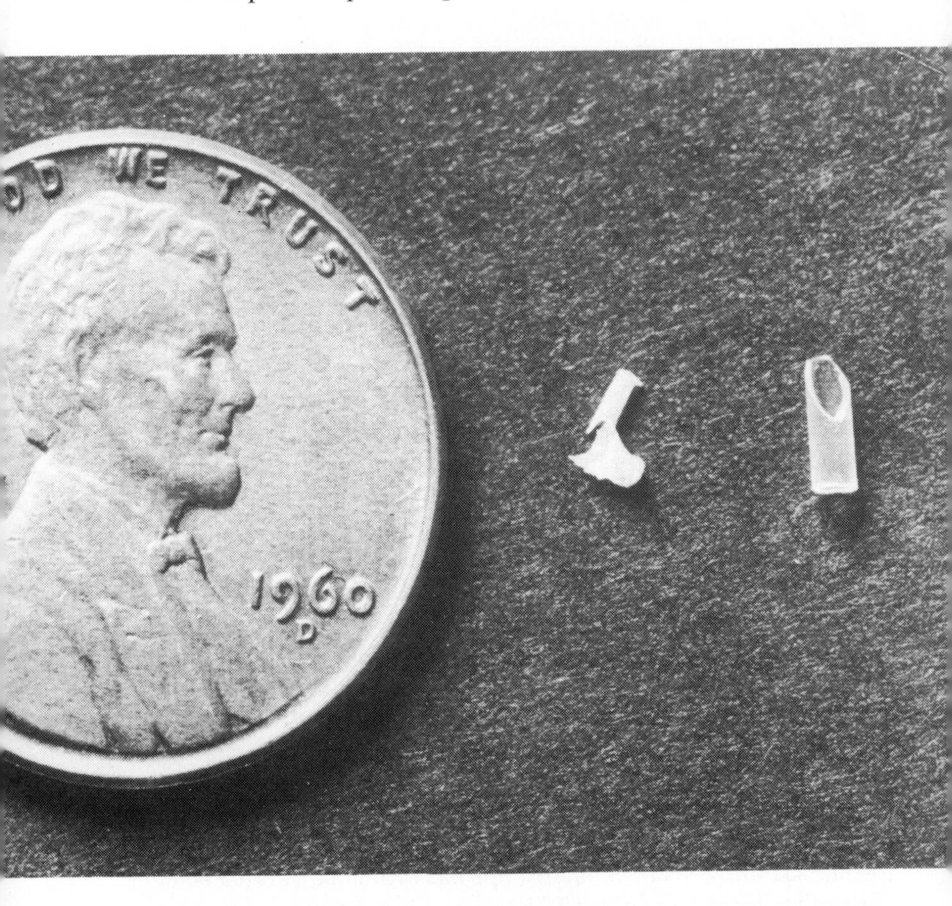

The microscope in surgery. Deafness due to disease of the stapes (small bone in the centre) can be cured by replacing the bone with a small polythene prosthesis. The cent (approximately same size as the English 1p piece) gives scale.

Moon dust reveals radio-activity of past ages. This small particle, seen with the electron microscope, contains numerous microscopic tunnels. They are 'scars' left by the passage of high-energy cosmic ray particles through the moon-dust. The tracks—'fossilized' radiation indicators—give information about past levels of radiation. Micrograph by Gareth Thomas, University of California, Berkeley.

materials—such as carbon-fibre, whose first widespread application was to have been in the fan-blades of the RB 211 jet engine—depend on microscopic structures about which we would otherwise know little. Many of the problems of wear, corrosion and performance are solved daily by branches of microscopy that the layman would not ordinarily hear about.

It is not only an awareness of microstructure that has influenced the development of modern industry. Some surprising discoveries have centred on micro-organisms and their ability to disrupt industrial processes—and these are new to most industrialists.

The understanding of the existence of our 'germ' neighbours can throw up new interpretations of a whole range of problems. For example, fuel sludging and corrosion has long been a problem in aircraft design and maintenance. Now we know the root cause. It is not primarily chemical, physical or an electrical disturbance set up by inorganic impurities. Microscopic fungi, as single-celled organisms, can actually live in the fuel—using it as a food material—wherever droplets of water are present, and it is they which have clogged fuel jets and caused a degeneration in the fuel itself. This discovery was of vital importance for the aircraft industry. Concorde, for instance, has been modified and redesigned with these microscopic organisms in mind.

Microbes of this kind have been found in the equipment at metal refinement plants. They are common in the rolling oil of steelworks. Left on the surface of sheet steel, they seem able to resist the cleansing treatments and can remain in position even after the product has been painted.

Many of the blisters and rust spots found in motor-car finishes and on domestic equipment such as refrigerators are now known to have been due to these micro-organisms. In attempts to eliminate them from the oil-ways in steelworks, several plants have been closed down and scalded with steam—surely the first examples of 'heat fumigation' against these germs of heavy industry.

But organisms that can exist in oily materials need not only produce harmful side-effects. The propensity can be harnessed, and made to work *for* industry, rather than against it.

Diesel fuel, for instance, has always been liable to thicken in cold weather until it may become unusable. It is known that this

The microscope in industry. Synthetic diamonds produced from granular graphite by scientists at the General Electric Company. This microscopic view including the point of a pencil gives an indication of their size.

is due to waxes dissolved in the oil itself. But several species of organism can destroy waxes, by using them as a foodstuff. And using this principle it is now possible to purify the fuel oil biologically. The organisms—a yeast, *Candida lipolytica*—are put to grow in the fuel and when they have eliminated the wax (and so purified the product, making it economically more attractive) they are removed as a brownish sludge which is rich in protein and B-vitamins, and which can be used as cattle-cake.

It has been suggested that the treatment of only three per cent of the world's crude oil production each year would not only purify the product but would provide enough protein by-product to equal the entire annual output of the world's fisheries. This could overcome the bulk of the foodstuff deficiency that we see in the under-developed nations.

There is very little new in the idea, even in principle; we have accepted this form of protein for years. A whole range of 'yeast extract' preparations are available, and protein concentrates such as Marmite and Yeastrel are popular. The yeast from which they are made, *Saccharomyces cerevisiae*, is a near relative of *C. lipolytica*, and we have accepted these foods without question. Clearly it is basically a matter of public attitudes and education.

There is now a growing awareness that micro-organisms might provide solutions for industrial problems; we could harness them as aids, rather than suffering from their effects as adversaries. It is brewing which provides the key. In this process *S. cerevisiae* is the yeast which ferments a brew of organic materials, and converts much of the sugar and starch in the mix into alcohol.

Other fungi, indeed other types of organism, can produce a whole range of important materials from an endless variety of mash mixtures. The enzymes that attack proteinaceous dirt in 'biologically-active' detergents are obtained by the mass-production of micro-organisms. Other bacterial species can ferment sewage residues to produce a harmless, powdery fertilizer and pure drinking water as end products—and the methane gas that is evolved in the process (identical to natural gas from the North Sea) can provide heat and light for the treatment plant.

So the cycle of microscopy has almost turned the full circle. We have gone from the sixteenth-century speculations about infectivity to the seventeenth century's first glimpses of microscopic organisms; through the next two hundred years of discovery and debate until the dawn of realization in the nineteenth century. The public were suddenly aware that there were unseen 'creatures' around that could kill them or make them ill.

Now, at last, we could mature into the acceptance of microbial life as an integral part of our civilization. This change and development of our attitudes could allow us, at last, to see micro-organisms as potential allies, as minute, self-regulating powerhouses that our industry can harness in the battle against waste, the search for new materials, and for harmless sources of technological energy.

It has taken a long time. In fact it is incredible to reflect on the antiquity of those primitive ideas about infection, the sophistication of Fracastoro's 'seminaria' thesis of the sixteenth century, in the light of discovery in the following century of the 'seminaria' or 'germs' themselves. Here—obviously—were the organisms that had been postulated. Yet no-one put two and two together, and the germ theory of disease continued to await discovery. Even earlier this century there were distinguished commentators who wrote of the role of 'fresh air' as a therapeutic agency; and it is still possible to find the belief that bad air or a foul smell can produce illness; so long it took for the ideas to gain credence.

It was not until 1876 that the brilliant mind of Robert Koch, then thirty-four and carrying out research at Wollstein, Germany, was able to demonstrate that a specific bacillus could cause a specific disease—anthrax. In those following few years the whole maze of the earlier work must have tumbled into place with a dramatic fervour of rare intensity. And the fact that the whole field had been so thoroughly explored beforehand must make it seem strange indeed that the evidence of those pioneers took so long to gel into reality.

Yet still we have so far to go.

The understanding of the nature of disease organisms has given us a vast range of vaccines and therapeutic drugs which between them have conquered or controlled many of the age-old plagues of our species. But the surface of the problem has hardly been scratched.

Still we cannot successfully treat *any* virus disease. People can be immunized against poliomyelitis, for instance, but if a patient does contract the disease there is no antidote that can eliminate the virus from his body. We can alleviate some of his symptoms; put him into a respirator if his breathing fails; but after that it is up to him. A vast outbreak could result from the escape of a lethal virus, and untold damage could be done to our civilization through the incalculable, tragic casualties that would result. Some viruses that have recently been documented for the first time have a unique pathogenicity towards mankind, and research workers have lost their lives in the cause of advancing this hazardous but important branch of research. It was in April 1971 that—quite by accident—a new virus was developed that causes a rapidly infectious cancer in human tissue cultures.

Yet, whilst the hypothetical menace of a 'super-virus' remains merely a possible threat for the future, we must realize that—whether virus or bacteria—we still know almost nothing about the mechanisms of infectious disease. How do these organisms make us ill? In the main we have no real idea of the nature of disease, nor how infections are actually caused. The self-perpetuating nature of bacteria, which are potentially immortal since they can avoid the ageing processes that afflict higher organisms, are another matter we have yet to understand. Indeed there is far more still to be discovered than we can possibly grasp—our investigations have, in truth, hardly started.

Quite apart from the ignorance that surrounds many of the ways of micro-organisms it is suddenly becoming apparent that there are whole new types of living thing that—even a couple of years ago—we did not suspect.

We are beginning to realize that there are 'slow viruses', a strange and hitherto totally unsuspected group of pathogens. There is evidence of their infectivity, but to date they have not been observed with the electron microscope, nor have they been cultured in the laboratory. They seem able to withstand high temperatures that would inactivate any ordinary virus, and they are immune to the effects of disinfectants: moreover they do not provoke the typical cell response that conventional viruses do, so that routine tests will not reveal their presence. They may turn out to be viruses of a kind, but perhaps they are genetic units that are free instead of fixed. Possibly they will turn out to be a simple form of entity at the border-lines of life. They just might account for a range of unexplained human conditions—but as yet we do not know.

In the microbial world there are the mycoplasmas. These strange organisms, which have for years been known as pleuro-pneumonia-like organisms (PPLO's for short) now seem to be an entirely new group of organisms, not bacteria, not fungi; almost like simple single-celled animals related to the protozoa. They have no true cell wall and have been observed to break into

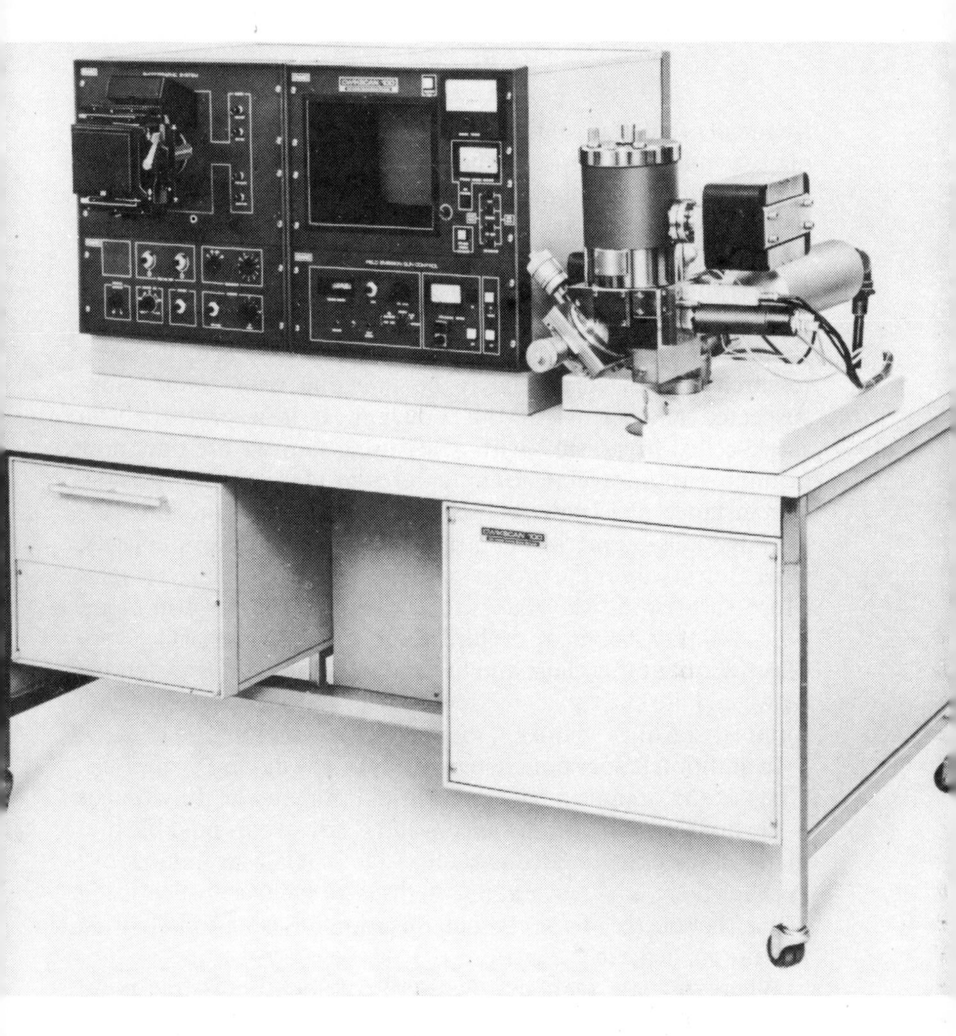

The 'Cwikscan-100' is the world's first field-emission scanning electron microscope to enter production on a commercial basis. Designed by the California company of Coates & Welter, it produces an electron beam from a tungsten tip that operates at room temperatures, and at a low voltage. The beam passes through an extraction anode and is accelerated by the application of a variable voltage of 400-20,000 volts to a second anode. The resolution is stated to be 100Å, but a second version announced towards the end of 1972 attained 50Å. The television display (centre) provides a continuous monitoring facility with a resolution of 1050 lines—almost twice that of a domestic UHF television receiver.

fragments that can continue growing. These tiny particles seem almost similar to viruses. They have been isolated from widespread samples of animal and human tissues, and some workers have suggested that they may explain some of the diseases in which the body seems to 'turn on itself' (arthritis, for example). Here is another vast new field of research—an unexplored galaxy in itself.

Another example is the genus *Toxoplasma* which, if recent research proves correct, may be infecting your tissues—unsuspected and silent—at this moment. It is a protozoan—a single-celled organism—with a life cycle that we are only now beginning to unravel. Certain features of it are most surprising. For instance, it is now realised that all stages in the life cycle are infectious—a marked contrast to other protozoan parasites, where infectivity is the property of one or two specialized stages. More surprising still, *Toxoplasma* can infect any warm-blood host, whether mammal or bird. Parasites are, as a rule, very selective about their host and in some cases are non-transferable between quite closely related species. But *Toxoplasma* seems able to infect starlings, monkey, rats and mice—and man.

In addition it does not seem to produce any disease symptoms. Cysts of the organism have been found in many healthy tissues—the human brain, for instance—and it now seems possible that many of us carry it in our tissues. Or is it all so innocuous? Perhaps there is a connection with diseases such as bronchitis, in which the spectre of a protozoan infection has occasionally tried to raise its head in the past.

Whereas a few years ago we were confidently existing in an academic climate that lulled us into apathetic acceptance of the view that we knew all the microbes there were, vast new vistas are beginning to appear. We can now discern countless new bacteria, fungi and viruses awaiting investigation; undreamed-of groups of protozoa, indeed whole classifications of life forms that—a few years ago—were never envisaged.

It is now fashionistic to condemn much of the early research work as premature, ill-judged or naïve: yet how far have we progressed when we are grasping for atoms and molecular lattices just as our predecessors were searching for the homunculus in the tiny head of a sperm? Only hindsight, so convenient to the commentator yet so unattainable by the pioneer, gives us the spurious confidence we need to scoff; and in other fields we are being as shortsighted ourselves.

The past mania for taxonomy—for the classification of living things in strict categories—has been seen, at last, as a menace to science. Bacteria, for instance, have been shown to exhibit the remarkable propensity to exchange genetic units; in this way a concept of a fixed species has sometimes been replaced by that of a continuous spectrum of types with endless genetic combinations possible. Forms of classification that were proposed only a matter of a handful of years back are now being disclaimed and withdrawn by their very proposers.

Indeed—such is the nature of fashionism, and so valid is it as a concept in the study of scientific progress—the pointer has swung so far back that it is almost off the dial altogether. Classification is now seen almost as a ludicrous aim in itself. Microbiologists are found who argue against the very existence of taxonomy. This is, patently, as absurd. The anthrax bacillus is still that; a recognizable entity with individual characteristics. So is the bacterium of tuberculosis, even though it has strains of varying degrees of pathogenicity. So is the proteus organism, and the many other recognizable entities. All that has happened is that we are recognizing the falsity of over-classification—and where does that, logically, leave us? Why, back in the late nineteenth century, when organisms were spoken of in larger, ill-defined assemblies.

Truly the wheel has turned full circle!

At each stage of microscopical awareness the men involved have felt that here, at last, was the final insight, the ultimate revelation. But like walking up a steep yet slowly levelling mountain, at each step a new horizon appears above and in front of us. We still have so far to go. The final conceptual breakthrough to a unified understanding of microscopic living organisms—a kind of biological theory of relativity—is still not even a dream, let alone tangible, attainable reality.

It is with condescension that we look back on earlier views on contagion.

RING A RING OF ROSES chant little children, in unknowing allusion to the skin rash, rosy-red, that marked the plague of the Middle Ages.

A POCKET FULL OF POSIES is the chanted reference to the nosegay of sweet-smelling herbs which—since it countered the evil odours that were believed to cause disease—was thought to protect against the illness.

ATISHOO, ATISHOO goes the rhyme as it mimics the symptom of sneezing so characteristic of the illness as it progresses.

WE ALL FALL DOWN ends the chant as the illness takes its fatal course.

There are remnants of these age-old beliefs in our language habits, too. Malaria, remember, means—literally—bad air. Influenza derives from the Italian, and reminds us it was once believed to be due to extraneous evil 'influences'. And it is still popular to say 'bless you!' after a sneeze in distant memory of the plague itself.

And how different is it now? Disinfectants are sold according to their 'clean-smelling' propensity. Once the magic ingredient was chlorophyll; then it was pinewood. Now it is lemons that litter the detergent and health advertisements. Yet they are nothing more than symbols of superstition, modern counterparts of fragrant nosegays and lucky charms. What will it be in a few years' time? Essence of orange? Tincture of tamarisk? And how long is it before we return to bat's-wing brooches or necklaces made from finger-bones—surely at this rate it cannot be long?

Certainly amongst older people in England it is still popular to believe that the odour of decay can cause disease. Not long ago I was told, by an elderly friend, that the smell from a neglected nest of kittens, abandoned in an outhouse, might give rise to diphtheria. In the industrial North a sweaty sock is still used as a remedy against sore throats, and goose-grease is employed in the Swansea Valley and elsewhere in rural Wales as a 'cure' for them.

And is it so very different in the laboratory? Still a bunsen burner is kept burning alongside culture plates during transfer operations. It is useful, certainly, for flaming the loop—for sterilizing the platinum wire in the flame—but we all are taught, and in our turn teach, that the flame helps to prevent contamination in some way. It does not. If anything, the air currents set up by it are more likely to waft extraneous organisms in to the area rather than exclude them. But perhaps the cleansing flame is seen, subconsciously, as a guard against microscopic organisms—a medieval relic, in cultural terms; a symbol of security.

The 'blind spot' extends further.

Still there is no legislation covering micro-organisms, even those that are of high pathogenicity towards the human host. The strict laws against radio-isotopes, drugs, toxic chemicals, insecticides and the rest, even controlling specified plant pathogens, are not mirrored in controls of any kind on the use of microbes in research or in the unfettered field of manufacturing industry. (See Appendix.)

I doubt whether our dyed-in-the-wool attitudes towards new discoveries have altered fundamentally, either. A recent paper by a leading specialist in the study of protozoan parasites, dealing with some unexpected but well-documented findings, was rejected by two leading journals; one because the work was 'too preliminary', the other because it was 'too out-dated'. That is an affair laughably reminiscent of the seventeenth century.

There is still a trace of belief in spontaneous generation too, so long after Spallanzani and Redi; Alan Dale in his *Introduction to Social Biology* states that—as a boy—he used to join with playmates scattering horsehairs into the river Meese, believing them to change into eels. The elvers that arrived each year seemed to confirm the superstition!

And how much does hygiene mean to us when, as a London professor reported in 1972, a surgeon drops a bone graft on the theatre floor—and wipes it on his apron before proceeding? What grasp of basic principles are shown when a central-heating company in London issues a press statement about a biocide which, added to the water in a heating system, kills bacteria 'that might prove to be a health hazard'?

Bacteria and the other micro-organisms are almost universally thought to be harmful in the public mind. Though they are so important, so vital, and so ubiquitous, the public have no idea what they look like, let alone what they do. The specialist who diagnoses a microbiological problem in industry and prescribes a biocide to control the organisms is, in fact, working in a way that is wonderfully simple to understand, fundamental and elementary to teach. Yet the industrialist is content to view him and his little case of samples like a witch-doctor, an alchemist, or a magician. Members of the public who have known about the rings around Saturn, the craters on the Moon, the intricacies of electronics, have no idea what a cell looks like, or what bacteria are. Yet these organisms are of far greater import.

All this alongside the ineptitude of many academic attitudes to the subject makes us realize that, for all the progress we have accomplished, we still retain our corporate 'blind spot' for the subject as a whole. The ever-receding skyline has yet to reveal the distance of the true horizon. When we do reach the summit of the rise it will be to realize that it is merely part of the foot-hills, with further, taller ranges still to seek, survey and conquer.

And what of the over-sold claims made each time some new development occurs? Are we really near the discovery of the 'key' to the mystery of life, the quest of mortal man for immortality, or an ultimate insight into the hidden universe? Perhaps.

Yet a key can do no more than unlock a door. That does but little to reveal the complexities that exist inside.

And it does nothing to help us understand.

Appendix

BIOHAZARD LAW

THE complete lack of legislation covering the handling of microorganisms referred to on p. 199 could be rectified by measures similar to those covering the use of radio-isotypes, drugs, poisons and the rest. There are many reasons to suppose that the lack of legal controls could become a danger to society.[1] The first publication of such a proposal[2] was supported actively by the press[3,4] and since then several episodes have strengthened the case. In the spring of 1973, for example, children were found to have been playing with containers of infectious material, a risk which the proposed legislation would eliminate. This aspect was discussed in *The Times*,[5] and subsequently the proposal was discussed at length in a leader column published in the same newspaper.[6]

The suggestion has already been published, in a textbook published in March 1973, that 'it is high time there were legal restraints placed on the conduct of research in microbiological laboratories, and it is still regrettably true that there are episodes of infection resulting from the deficient techniques used in many establishments'.[7] One month later there occurred the tragically fatal outbreak of smallpox from a research laboratory, which is clearly a reminder of the hazards attached to the experimental use of pathogens. It would be indeed unfortunate if the 'blind spot' we have shown towards microbial life for so many centuries (p. 193 *et seq.*) extended to a failure to regularize laboratory research. The proposed legislation, which is now being discussed by several Members of Parliament, is outlined below.

[1] Ford, B. J.: 'No Law for Bacteria' (*Observer*, 1 August, 1971).
[2] ——: 'No Legal Control of Biological Hazards' (*New Law Journal* 121 (5511) 823).
[3] Anon: 'Disease "risk" in Labs' (*The Guardian*, 17 September, 1971).
[4] Berlins, M. and Wright, P.: Call for Law to Control Laboratory Poisons (*The Times*, 17 September, 1971).
[5] Wright, P.: 'Absence of Regulations for Microbes is Risk to Public Health, Expert says' (*The Times*, 11 April, 1973).
[6] Leader column 'A Case for Much Wider Inquiry' (*The Times*, 18 April, 1973).
[7] Ford, B. J.: *The Optical Microscope Manual* (David and Charles, 1973), p. 196.

THE PROPOSALS

SCHEDULING OF ORGANISMS. Micro-organisms and viruses would be classified according to two sets of criteria, designed to identify them as of *high* and *low* risk respectively.

SCHEDULE A organisms would be the most dangerous—i.e., pathogens that are *not* normally present in the environment and which are known to cause epidemic outbreaks when they *are* present. The causative factors of smallpox, plague, tuberculosis, etc. are included under this heading.

SCHEDULE B would encompass pathogens that are known to cause disease in man, but which are often found in the environment, and which are hazardous only when cultured, or when present in significant amounts. Bacteria such as *Staphylococcus*, *Streptococcus* and some species of *Salmonella* are examples.

REGISTRATION. The culture of pathogens would be restricted to properly competent individuals, or technicians under their supervision, and holders of Schedule A pathogens would be registered. This would facilitate research, and would guard against the production of cultures in unsupervised circumstances.

SAFETY STANDARDS. The prohibition of certain procedures at present widely used in laboratories, yet known to be dangerous, would be facilitated under this provision. The use of sharp hypodermic needles to transfer hepatitis serum is one example; the use of unplugged mouth pipettes is another. Inadequate conditions for the disposal of infective material, such as the pouring of culture media down a sink, or the discarding of containers with other refuse, would be prohibited.

CO-ORDINATION. As a result the co-ordination of standards, the dissemination of new findings, and the protection of both research worker and public would be greatly facilitated. The enactment of such legislation (particularly in view of the laws covering isotopes, drugs, poisons, etc) is long overdue.

Bibliography

BRADBURY, S.: *The Evolution of the Microscope* (Pergamon Press, 1967).

BROCK, T. (ed.): *Milestones in Microbiology* (Prentice-Hall, 1961).

BULLOCH, W.: *History of Bacteriology* (Oxford University Press, 1938; Athlone Press, 1961).

CLAY, R. S. and COURT, T. H.: *History of the Microscope* (Griffin, 1932).

DOETSCH, R. N. (ed.): *Microbiology* (Rutgers University Press, 1960).

FORD, BRIAN J.: *Microbiology and Food* (Northwood, 1970).

——: *The Optical Microscope Manual: Past and Present Uses and Techniques* (David and Charles, 1973).

MAJOR, R. H.: *A History of Medicine* (Blackwell, 1955).

ROSEBOOM, M.: *Microscopium* (Leiden, Holland).

ROUSSEAU, P.: *Histoire de la Science* (Lib. Arthème Fayard, Paris, 1945).

Index of Names

Index of Names

Subject Index